The Multiwavelength Atlas of Galaxies

Since the radio signature of our own Milky Way was detected in 1931, galaxies have been observed from ultra-high-energy gamma rays to long-wavelength radio waves, providing fundamental insights into their formation, evolution and structural components.

Unveiling the secrets of some of the best-observed galaxies, this atlas contains over 250 full-color images spanning the whole electromagnetic spectrum. The accompanying text explains why we see the component stars, gas and dust through different radiation processes, and describes the telescopes and instruments used.

This atlas is a valuable reference resource on galaxies for students seeking an overview of multiwavelength observations and what they tell us, and for researchers needing detailed summaries of individual galaxies. An accompaying website, hosted by the author, contains slide shows of the galaxies covered in this book. This is available at www.cambridge.org/9780521620628.

Glen Mackie is a Lecturer in the Centre for Astrophysics and Supercomputing and Assistant Coordinator of Swinburne Astronomy Online (SAO) at Swinburne University of Technology. His research interests include galaxy formation and evolution in the rich cluster environment, multiwavelength properties of galaxies, galaxy interactions and mergers, astronomy education and history.

The Multiwavelength Atlas of Galaxies

Glen Mackie
Swinburne University of Technology

CAMBRIDGE UNIVERSITY PRESS
Cambridge, New York, Melbourne, Madrid, Cape Town, Singapore,
São Paulo, Delhi, Dubai, Tokyo, Mexico City

Cambridge University Press
The Edinburgh Building, Cambridge CB2 8RU, UK

Published in the United States of America by Cambridge University Press, New York

www.cambridge.org
Information on this title: www.cambridge.org/9780521620628

First published 2011

Printed in the United Kingdom at the University Press, Cambridge

A catalog record for this publication is available from the British Library

Library of Congress Cataloging-in-Publication Data

Mackie, Glen H., 1961–
The multiwavelength atlas of galaxies / Glen H. Mackie.
    p.   cm.
ISBN 978-0-521-62062-8 (Hardback)
1. Galaxies–Atlases.   2. Galaxies–Observations.   I. Title.
QB856.M33 2010
523.1′12–dc22   2010033387

ISBN 978-0-521-62062-8 Hardback

Additional resources for this publication at www.cambridge.org/9780521620628

For Corinne, with love.
She outshines even these brilliant galaxies.

Contents

Appendices and references   219

Preface

In the early 1930s Karl Jansky[1] detected the center of our Galaxy in radio waves at a wavelength of 14.6 m or a frequency of 20.5 MHz.

Until this time all observations of galaxies had been made in the optical region, which has radiation with wavelengths between 330 and 800 nanometers. Galaxies that can be seen by eye, appearing as faint, cloud-like objects, are the two Magellanic Clouds (named after the Portuguese explorer Ferdinand Magellan) in the southern hemisphere and the Andromeda Galaxy (NGC 224 or Messier 31) in the northern hemisphere. Observers with very good eyesight may also detect NGC 598/M 33, in the constellation Triangulum. We also get a myopic view of nearby stars (about 6000 within a few thousand light-years) and a wide-field view of diffuse light, "the Milky Way", from distant stars, near the plane of our own Galaxy.

Since the early seventeenth century, telescopes have provided higher resolution observations to fainter brightness levels than possible with the eye. In 2009 the world celebrated the International Year of Astronomy (IYA), as sanctioned by the United Nations and directed by the International Astronomical Union partly to celebrate 400 years of telescopic observations since the initial endeavors of Thomas Harriot and Galileo Galilei.

In terms of effective galaxy research tools, we had to wait until the mid-nineteenth century for large telescopes to be built. Observations by eye, photography and by spectroscopy all played vital roles. The Andromeda Galaxy, and others, became known as "spiral nebulae" and came under close scrutiny. The nature and content of the spiral nebulae were increasingly debated until the mid-1920s when their true distances were revealed. The spiral nebulae were not part of our Galaxy, which many had thought constituted the Universe. They were separate, distant galaxies in their own

right. Almost overnight, the true nature of the Universe was discovered. Our Galaxy was one of billions.

Since Jansky's breakthrough in the early 1930s astronomers interested in studying the formation and evolution of galaxies have rapidly ventured into other regions of the electromagnetic spectrum. In 1951 Ewen and Purcell discovered 21 cm radio emission from interstellar hydrogen, as predicted by Van de Hulst in 1945. In 1961 the satellite Explorer 11 detected 22 events from high-energy gamma rays (and 22,000 events due to charged cosmic rays!). The first extrasolar X-ray source was detected in 1962, using a sounding rocket, by the X-ray astronomy group at American Science and Engineering. In the late 1960s Neugebauer and Leighton completed the first major near-infrared survey of the sky. From 1968 onwards Orbiting Astronomical Observatory (OAO) satellites provided the first detections of sources in the ultraviolet. In 1987 the James Clerk Maxwell Telescope (JCMT) began dedicated observations in the submillimeter region. The optical and near-infrared regions have also benefited by advances in instrumentation and active and adaptive optics on the latest generation of very large ground-based telescopes. Near-diffraction-limited observing has been achieved by placing telescopes above the Earth's atmosphere (e.g. the Hubble Space Telescope since 1990).

Radiation from all regions of the electromagnetic spectrum has now been detected from galaxies. These observations have utilized telescopes at ground-based observatories (many located at high-altitude mountain sites), telescopes in aircraft (>10 km altitude), detectors on balloons that voyaged to the upper atmosphere (30–40 km altitude), instrument payloads on rockets that reached space, observatory satellites in Earth and solar orbit, as well as instruments aboard planetary missions journeying through the Solar System.

---

1 K.G. Jansky, 1933, Electrical disturbances apparently of extraterrestrial origin, *Proceedings of the IRE*, **21**, 1387; K.G. Jansky, 1933, Radio waves from outside the solar system, *Nature*, **132**, 66.

This atlas is a compendium of galaxy images spanning the gamma ray, X-ray, ultraviolet, optical, infrared, submillimeter and radio regions of the electromagnetic spectrum. Explanatory text describes how different radiation is produced, which objects (i.e. cold, warm or hot gas, dust, stars, particles, atoms and molecules) it originates from, and what types of telescopes are used to detect it. The galaxies are divided into categories of normal, interacting, merging, starburst and active, though many have been classified across one or more categories.

The purpose of this atlas is to display and describe some of the best multiwavelength images of galaxies. The images originate from a variety of telescopes, instruments and detectors, and therefore possess wide ranges of signal-to-noise,[2] angular resolution, sampling or pixel sizes and fields of view.

The atlas consists of four parts. The first part introduces galaxies, discusses the atlas sample, categories of galaxies and properties of the sample. Part 2 (page 29) describes the various origins of radiation emitted by galaxies and presents some important galaxy research topics that use multiwavelength observations. Part 3 (page 55) discusses multiwavelength radiation from the Galactic center and finishes with all-sky multiwavelength images of our Galaxy. Part 4 (page 67) comprises the multiwavelength atlas images with scientific descriptions from the literature. The book concludes (page 219) with appendices describing telescopes and instruments, image sources, technical descriptions of each image, a cross-reference list of galaxies, and example plots of spectral energy distributions. General text references and atlas galaxy references are also given.

The images presented will help the general reader appreciate the incredible information content of multiwavelength observations without needing an extensive scientific or mathematical background. The accompanying text provides specific discussion of galaxy properties, and astronomical radiation origins and processes useful for those wanting more astrophysics content. Research summaries of individual atlas galaxies are given for those interested in more detailed multiwavelength information.

An accompanying website
http://astronomy.swin.edu.au/~gmackie/MAG
exists. I would appreciate correspondence from researchers who may have relevant information and/or images.

Glen Mackie,
gmackie@swin.edu.au,
Centre for Astrophysics and Supercomputing,
Swinburne University of Technology,
December, 2009, IYA

---

2 Signal-to-noise is a measure of the amount of detected radiation from a source above the inherent radiation or noise of the detector and background signal.

Acknowledgements

This atlas has come about due to the generosity of many astronomers that have kindly provided their data and information about their observations. I would like to thank P. Amram, P. Barthel, R. Beck, L. Blitz, M. Birkinshaw, G. Bothun, G. Bower, A. Bridle, U. Briel, E. Brinks, M. Burton, H. Bushouse, B. Canzian, C. Carilli, J. Chapman, P. Choi, J. Condon, P. Cox, P. Crane, T. Dame, R. Diehl, S. Digel, S. Döbereiner, M. Dopita, N. Duric, E. Dwek, M. Ehle, G. Fabbiano, E. Feigelson, L. Ferrarese, D. Finkbeiner, K. Freeman, P. Guhathakurta, R. Genzel, M. Haas, D. Hartmann, R. Haynes, D. Heeschen, J. P. Henry, J. Hibbard, H. Hippelein, S. Holmes, J. C. Howk, J. Holtzman, Hubble Heritage Team, A. Hughes, E. Hummel, J. Hutchings, G. Jacoby, S. Jorsater, N. Junkes, P. Kahabka, J. Kamphuis, W. Keel, D. Kelly, N. Killeen, U. Klein, B. Koribalski, K. Kraemer, M. Kramer, M. Krause, R. Laing, S. Larsen, D. Leisawitz, G. Lewis, L. Lindegren, Local Group Survey Team, B. Madore, P. Massey, A. Mattingly, D. Maoz, A. Marlowe, A. McConnachie, S. McGaugh, K. McLeod, MCELS Team, D. Meier, K. M. Merrill, G. Meurer, C. Mundell, N. Neininger, R. Norris, U. Oberlack, K. Olsen, A. Pedlar, R. Perley, W. Pietsch, J. Pinkney, P. Planesas, R. Pogge, S. Points, M. Putman, R. Rand, S. Raychaudhury, T. Rector, M. Regan, W. Rice, G. Rieke, M. Rieke, A. Rots, M. Rubio, S. Ryder, B. Savage, E. Seaquist, E. Schlegel, S. Schoofs, N. Scoville, E. Shaya, S. Shostak, C. Smith, S. Snowden, W. Sparks, S. A. Stanford, L. Staveley-Smith, R. Supper, R. Sutherland, P. Thaddeus, R. Tilanus, S. Tingay, G. Trinchieri, J. Trümper, B. Tully, J. Turner, J. Ulvestad, E. Valentijn, J. van der Hulst, G. van Moorsel, L. Vigroux, A. Vogler, W. Walsh, R. Walterbos, B. Whitmore, R. Wielebinski, A. Wilson, F. Winkler, T. Wong, J. Wrobel, C. M. Young, and H. Zimmermann.

Credits for atlas image observers and data owners are also summarized in Appendix C (page 224).

Seth Digel (GSFC) was of great assistance in providing many of the Astrophysics Data Facility all-sky images of the Galaxy and reformatting several images. Karen Smale (LHEA), Kathy Lestition (Chandra X-ray Observatory Center), Dan Brocious (Smithsonian Institution, Whipple Observatory), Catherine Ishida (Subaru Telescope) and support staff at STScI helped with my many enquiries. I worked on the beginnings of this atlas whilst visiting the Anglo-Australian Observatory and I thank the then Director, Prof. Brian Boyle, for his hospitality.

I would also like to thank the editorial staff at Cambridge University Press, especially Adam Black for his initial interest in the atlas, and Vince Higgs and Lindsay Barnes for their expert guidance to get it to completion.

The construction (*construction* tends to more adequately describe the process of producing this atlas than writing!) and presentation of this work were greatly improved due to the keen eye of my wife, Corinne. She quizzed me on many statements during enthusiastic readings of the chapters. I thank her for wonderful support and encouragement.

I hope this atlas represents a tribute to the rapid progress made in multiwavelength observations of galaxies and credit is due to all astronomers involved in this type of research. It was always heartening to hear support from astronomers who corresponded with me about the atlas during its construction, and who were extremely generous with their time and knowledge: many thanks to Rainer Beck, John Hibbard, Annie Hughes, Phil Massey, and Sean Points.

I received valuable comments on drafts of this atlas from astronomers Chris Flynn, Virginia Kilborn, Bill Keel, Malcolm Longair, Emma Ryan-Weber and Steve Tingay which greatly improved the final product. I have also benefited from comments from anonymous referees appointed by Cambridge University Press. Any errors or omissions remaining are entirely my responsibility.

G.M.

PART 1

Galaxies

If galaxies did not exist we would have no difficulty in explaining the fact. WILLIAM SASLAW

GALAXIES are fundamental constituents of the Universe. They are groups of approximately 10^5–10^{13} stars[1] that are gravitationally bound and take part in the general expansion of the Universe. Galaxies have diameters ranging from 10,000 to 200,000 light-years[2] and they possess widely varying gas and interstellar dust contents. The distances between galaxies are typically 100–1000 times their diameters. They represent 10^8 overdensities above the mean stellar density of the local Universe. In total, galaxy masses range from 10^6 to 10^{14} solar mass ($M_\odot$).

Their component stars vary from ~0.1 $M_\odot$ brown dwarfs that do not undergo thermonuclear fusion, to rapidly evolving stars of at least ~50 $M_\odot$ and possibly as massive as 100 $M_\odot$. The stars' evolutionary state ranges from protostars undergoing contraction to begin thermonuclear reactions, to main-sequence dwarf stars that fuse hydrogen to helium in their cores, through to red giant stars with expansive gaseous atmospheres. The stellar end-products are white dwarfs, neutron stars and black holes. The specific evolutionary path of a star is governed by its mass. The most massive stars evolve over millions of years, whilst the lowest mass stars can evolve for billions of years.

At least 300,000 years after the Big Bang start of the Universe, structures that would become galaxies began to condense out of primordial hydrogen and helium. Current observational and theoretical studies of the formation and evolution of large-scale structure (groups, clusters and superclusters of galaxies) suggest that cold dark matter (CDM) is the predominant matter in the universe. Observations of individual galaxies suggest that CDM is the dominant matter in galaxies (see Section 2.5 for more information), making up more than 50% of a galaxy's mass. CDM, a theoretical massive, slow-moving particle (or particles), has not yet been detected. However, a CDM dominated universe would suggest that galaxies were built from the aggregation of smaller structures, in a sort of "bottom-up" construction approach. In fact the closer galaxies are detected to the time of the Big Bang, the stronger this argument becomes.

From about 300,000 years after the Big Bang, the attractive force of gravity was in control and dictated that the first galaxies formed within 1 billion years (Gyr). Gravity increased gas densities and temperatures until physical conditions allowed the first stars to form. Elements heavier in atomic weight than hydrogen and helium were then made by thermonuclear fusion inside the first (massive) stars, and expelled into the interstellar medium (ISM) by subsequent supernovae explosions. These explosions propagated shock waves through nearby interstellar gas causing gas densities to increase and new star formation to be initiated. Numerous cycles of star formation–supernovae explosions–star formation continued, utilizing the increasingly heavy element enriched ISM.

Enough time, about 14 Gyr, has now elapsed since the beginning of our Universe, to allow a stable system of planets rich in heavy elements to exist around a very average G dwarf star called the Sun. Our Sun is in the

Figure 1.1 (opposite)

Optical image of the barred spiral galaxy NGC 1365 in the Fornax cluster of galaxies.

Credit: SSRO/PROMPT and NOAO/AURA/NSF.

---

1 The exponent above 10 represents the number of zeros after 1; e.g. 10^9 is 1000,000,000 or 1 billion. If the exponent is negative, e.g. 10^{-2} then it is 1 divided by 10^2 or 1/100 = 0.01.

2 A light-year is the distance that light travels, at 2.9×10^5 kms^{-1}, in a year. It is approximately 9.46×10^{12} km.

outer suburbs of a dense disk of stars that has spiral arms of gas and dust[3] where stars preferentially form. A lower density, large spheroidal halo of predominantly older stars surrounds the Galaxy.[4] Our Galaxy contains at least 100 billion stars, and it is regarded as an average sized spiral. From our vantage point our Galaxy and a vast variety of other galaxies are observed.

The appearance of galaxies varies due to their intrinsic properties and the differing lines of sight (with differing gas and dust contents) through our Galaxy along which they are observed. These three-dimensional objects are also viewed projected on a two-dimensional sky surface or celestial sphere. Hence their angle of inclination to our line of sight greatly influences how an observer perceives them.

Galaxies can be broadly grouped into elliptical, spiral, lenticular and irregular classes. Ellipticals range from circular shaped to highly elongated or oval-like congregations of predominantly old, evolved stars. Spirals are Catherine wheel-like, with "fireworks" of young stars, dust and gas illuminating majestic spiral arms that can wrap around their nuclei many times. Stellar bars (elongated distributions of usually old stars) can be found in the centers of some spirals giving a spindle-like appearance (see the barred spiral NGC 1365 in Figure 1.1). The lenticulars (or S0 types) can be regarded as a visual link between ellipticals and spirals. They are disk- (like spirals) or lens-like though composed of predominantly old stellar populations lacking in dust and gas (like ellipticals). The irregulars, as their name suggests, lack coherent spiral or elliptical shapes. They contain regions rich in dust and gas and have active star-formation sites, yet are structurally quite amorphous.

1.1 Prehistory of galaxies

The term "galaxy" has its origin in the Greek *kiklos Galaxias* meaning milky circle, which is a reference to the diffuse, cloudy appearance of the Milky Way.

The discovery of galaxies as individual stellar systems scattered throughout the Universe took a long and arduous path. The Andromeda Galaxy, M 31, and the Large Magellanic Cloud had been recorded by Al-Sufi in his publications in the tenth century and were known to people of the Middle East. From the voyage of Ferdinand Magellan in 1519–1522 that took him to southern latitudes, the Large and Small Magellanic Clouds (Figure 1.2) became better known in Europe,[5] however they were not regularly known by these names until the nineteenth century.

In 1612 Simon Marius observed M 31 with a telescope and described its nebulous appearance as "the flame of a candle is seen through [a] transparent horn". In 1774 Charles Messier included the Andromeda Galaxy as the thirty-first entry (Messier 31) in his first catalogue (Messier 1784) of "nebulae" that also included other galaxies such as the satellite of M 31, the smaller M 32 and M 33 in Triangulum.

By the mid-eighteenth century the Englishman Thomas Wright proposed that the Sun was positioned away from the center of a slightly flattened, yet predominantly spherical system of stars. His ideas were published in his *An Original Theory or New Hypothesis of the Universe* (Wright 1750). Wright was one of the first to begin to understand the appearance of the Milky Way by discussing the role of an observer looking at a three-dimensional stellar system (their "universe").

The philosopher Immanuel Kant read about Wright's work but did not see the original diagrams. It was this omission that strangely helped Kant to hypothesize a disk of stars, which would naturally explain the appearance of the Milky Way in the sky. He then took the next step, amazing at the time, of proposing that similar "disks" of stars were scattered throughout the Universe as he was aware of observations of many "nebulous objects" detected around the sky. He published *The Universal Natural History and Theory of Heavens* (Kant 1755) though it was not distributed widely.

---

3 Interstellar dust grains make up about 1% of the ISM and are formed in the envelopes of late-evolved stars like red giants or carbon stars. They are much smaller than the dust on Earth, ranging from a few molecules to a diameter of 0.1 mm. They begin as carbon or silicate grains, and then accumulate additional atoms (e.g. H, O, C, N) to form icy mantles of water ice, methane, carbon monoxide and ammonia. All of this is encased in a sticky outer layer of molecules and simple organic compounds created through the interaction of the mantle with ultraviolet (UV) radiation. The grains are about the same size as the wavelength of blue light, meaning that they absorb and scatter UV and blue light much more efficiently than red light.

4 Throughout the atlas, the Milky Way, our Galaxy, will be denoted by an upper case G.

5 Amerigo Vespucci possibly mentioned the existence of the two Clouds and the dark Coalsack Nebula in a letter during his third voyage of 1503–1504 that took him to the south-east coast of South America. Andrea Corsali, an Italian serving on a Portuguese expedition, in a letter written in 1515 to Giuliano de Medici sketched both the Southern Cross and the two Clouds.

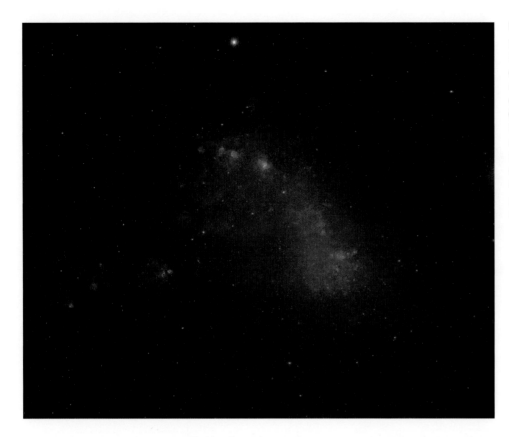

Figure 1.2

Optical image of the Small Magellanic Cloud. The image was constructed from more than 500 separate images, using green and red continuum, and narrowband [O III] (blue), [S II] (green), Hα (red) filters. Star light (green and red continuum) is suppresed to enhance the interstellar medium as indicated by the gaseous emission lines [O III], [S II] and Hα that occur in or near star-forming regions.

Credit: F. Winkler/Middlebury College, the MCELS Team and NOAO/AURA/NSF.

By 1761 Johann Heinrich Lambert, in a collection of essays, had independently suggested the same ideas, though history tends to concentrate more on the contribution of Kant. Kant and Lambert had, on the shoulders of Wright, launched the idea of "island universes" as separate entities. Wright, Kant and Lambert had many details wrong, yet they had together provided the starting points of the modern day view of the Galaxy, and the large-scale structure of the Universe.

Whilst a somewhat simplified theoretical basis had been laid, the observational efforts were hampered until large reflecting telescopes became available to try to resolve the structure of these "nebulous objects". In the last two decades of the eighteenth century, William Herschel, inspired by Charles Messier's catalogue of nebulae, tried to resolve these objects. Indeed for a time, he believed all such nebulae were star clusters. Interestingly, Messier had in fact observed the Virgo Cluster of galaxies, and Herschel had similarly seen the Coma Cluster. Determining the exact nature of "nebulae" was daunting as they contained objects as varied as globular clusters, supernova remnants, planetary nebulae, as well as galaxies.

The son of William, John Herschel, began to catalogue nebulae beginning in the 1840s (Herschel 1864) preceding Dreyer's New General Catalogue (Dreyer 1888) and two Index Catalogues of 1895 and 1908 (Dreyer 1895, 1908). John Herschel and William Parsons, Third Earl of Rosse, who built the 72 inch "Leviathan of Parsonstown" telescope, began a protracted debate about the stellar nature of the nebulae. Parson's sketches of spiral structure (e.g. the Whirlpool Galaxy, Messier 51, Parsons (1850), Figure 1.3) that cemented the term "spiral nebulae", are still widely recognized today.

Whilst the Leviathan of Parsonstown aided some observations, Parsons, erroneously, claimed to resolve the Orion Nebula into stars. Meanwhile, John Herschel began some of the first statistical studies of the all-sky distribution of the nebulae, and confirmed the Virgo Cluster and surrounding environs, as well as discovering the Pisces Supercluster:

> The general conclusion which may be drawn from this survey, however, is that the nebulous system is distinct from the sidereal, though involving, and perhaps, to a certain extent, intermixed with the latter.
>
> JOHN HERSCHEL

Despite Herschel's somewhat vague conclusion, the scientific chase was quickening. Astronomers adopted the term "galaxy" relatively quickly. In 1837, Duncan

Figure 1.3

Sketch of Messier 51 by William Parsons, Earl of Rosse. North is 90° clockwise (CW) from up.

Credit: From Parsons (1850).

Bradford published a popular account of astronomy that used the word "galaxy" for the Milky Way. By 1870 it seemed to have been generally accepted into the literature as various derivations of the term were being vigorously debated. In that same year the journal *Nature* (Evans 1870) stated:

> Mr. John Jeremiah states that "Heol y Gwynt" is the only proper Welsh name for the Milky Way. Such is far from being the case. I am acquainted with no less than nine other names, equally proper for that luminous appearance, such as y llwybr llaethog, y ffordd laeth, llwybr y gwynt, galaeth, eirianrod, crygeidwen, caer Gwydion, llwybr Olwen, and llwybr y mab afradlawn. Of these names, y llwybr llaethog and y ffordd laeth answer precisely to Milky Way; llwybr y gwynt (common enough in Carmarthenshire) is synonymous with heol y gwynt; galaeth (from llaeth, milk) corresponds with galaxy; eirianrod signifies a bright circle; and crygeidwen a white cluster.
>
> D. Silvan Evans

Whilst the Welsh had clearly cornered the market on Milky Way nomenclature and the residents of Carmarthenshire could readily describe their night sky, the physical nature of the nebulae had still not been determined. In the mid-1860s William Huggins used the new technique of spectroscopy to identify emission lines that showed many nebulae were gaseous. His initial observations of M 31 (Figure 1.4) and its nearby satellite galaxy M 32 showed no bright lines like those in the gaseous nebulae. They had continuous "star-like" spectra.

These became known as "white nebulae". About the same time Abbe (1867) studied John Herschel's 1864 catalogue and determined that Huggins, gaseous nebulae were distributed across the sky like star clusters and were thus part of the Milky Way whilst the other "white" nebulae were not. This essentially correct spatial separation was (unfortunately) not supported by many other astronomers of the time.

In 1885 a stellar transient was detected in M 31. Assuming a nearby distance for M 31 the event was regarded as a normal stellar event, a nova, and named Nova 1885. It is now realized that what occurred was the much more explosive supernova event, denoted S Andromedae, with its greater intrinsic brightness consistent with the modern cosmological distance of M 31.

Figure 1.4

Messier 31 in the ultraviolet: Galaxy Evolution Explorer's far-ultraviolet 135–175 nanometers (blue) and near-ultraviolet 175–280 nanometers (green). Infrared: Spitzer Space Telescope 24 μm (red).

Credit: NASA/JPL-Caltech/K. Gordon (University of Arizona), GALEX Science.

In the meantime, new clusters and even superclusters of nebulae were being noted by Stratonoff (Perseus–Pisces following on from Herschel), Easton and Reynolds. Notable clues were being collected. Spectroscopy had shown the stellar-like nature of the spiral or white nebulae, but were they nearby inside a single stellar system (our Galaxy) or outside at much larger distances? Spectroscopy also began to provide measurements of their velocities. Vesto Slipher at Lowell Observatory, beginning in 1912, detected large radial velocities in the nebulae from Doppler shifts[6] of various spectral lines. Within five years he had detected an average velocity of $\sim$570 km s^{-1} for about 30 nebulae compared to $\sim$20 km s^{-1} for stars (Slipher 1917). These velocities were far larger than could be explained if the nebulae were part of our Galaxy. The interpretation of the detected Doppler shifts and the subsequent inferred velocities was debated. Slipher had also improved upon Huggin's earlier spectra of M 31 and again noted a stellar spectrum by 1912. More discoveries of "novae" by Curtis and Ritchey in nebulae seemed to favor an extragalactic origin, though some were novae and others were supernovae.

There was still no great breakthrough but the foundations of discovery had been laid. In 1920 a topical "debate" between Harlow Shapley and Heber Curtis covered the nature of spiral nebulae. Curtis correctly argued that the nebulae were extragalactic but failed to convince the audience, whilst Shapley elegantly argued their local nature based on wrong conclusions. The answer, however, had already been found.

1.2 "It is worthy of notice"

Our present detailed knowledge of galaxies has been compiled over a relatively short time. It was not until the mid-1920s that spiral nebulae were proven beyond doubt to be separate galaxies in their own right. One astronomer made the breakthrough, and it began with photographic surveys of the Magellanic Clouds. Based on this data Henrietta Leavitt of Harvard College Observatory detected Cepheid[7] variable stars in the Magellanic Clouds (Figure 1.2). Beginning this work in 1905, with photographic plates taken by the Bruce telescope in Peru, the discovery can be traced to Leavitt (1908) who noticed a strange relationship in the properties of Cepheids in the Small Magellanic Cloud:

> It is worthy of notice that in table VI the brighter variables have the longer periods. H.S. LEAVITT

Leavitt had discovered that the star's period of variability (the time after a cyclical brightness change is repeated) was related to its mean brightness. By assuming that the width of the Small Magellanic Cloud was negligible compared to its distance away from us, which is essentially true, the apparent brightness of such stars would be related to their intrinsic brightnesses or luminosities. Hence the observed periods would directly correlate with the luminosity of the stars, allowing a distance to be calculated if the luminosity of Cepheids could be calculated.

---

6 The Doppler effect is the change in wavelength of radiation when the radiation source and observer are moving away or toward each other.

7 Cepheid variables are named after their prototype, δ Cephei.

If distances to local, nearby Cepheid variables with known periods could be determined, the period–luminosity relationship of Magellanic Cloud Cepheids could be calibrated and distances to the Clouds obtained. Leavitt had discovered the first useful cosmic yardstick! However, nature conspired to make such measurements very difficult. From 1913 to 1924 journal publications associated with Cepheid variables seemed to invariably attract the word "problem" in their titles and had piqued the interest of noted astronomers such as Stebbins, Duncan, Shapley, Hertzsprung, Cannon, Eddington, Adams, Jeans, van Maanen, Vogt, Luyten, Humason, Kapteyn and van Rhijn. This did not deter one of them, for as early as 1913 Ejnar Hertzsprung produced a first calibration of the period–luminosity relationship of Galactic Cepheids and derived an extremely large (at the time) distance of 10 kiloparsecs[8] for the Small Magellanic Cloud (SMC). Shapley, however, continued to argue that the Clouds were subsystems within our Galaxy. For comparison, the modern day distance to the SMC is 60 kpc.

Eventually a more accurate Cepheid period–luminosity calibration was derived and the apparent magnitudes[9] of Magellanic Cloud Cepheids were converted into absolute magnitudes, allowing a more accurate distance to the Clouds to be derived. This would definitively place them well outside the limits of our Galaxy.

In the meantime, other galaxies were being observed. From the early 1920s onwards John C. Duncan, Edwin Hubble and others did this and detected Cepheids in, amongst other Local Group galaxies, M 33 (Duncan 1922), M 31, IC 1613 and NGC 6822. At the same time, a theoretical treatment using the observed rotation of M 31 by Ernst Öpik (Öpik 1922) derived a large distance of 450 kpc. In the end it would be observations that comprehensively decided the matter.

A joint meeting of the American Astronomical Society and the American Association for the Advancement of Science was held in Washington, D.C. On New Year's

Figure 1.5 (opposite)

Messier 51, NGC 5194, the Whirlpool Galaxy in X-rays (purple), ultraviolet (blue), optical (green) and infrared (red). The companion galaxy NGC 5195 is also shown to the north.

Credit: X-ray: NASA/CXC/Wesleyan U./R. Kilgard et al.; UV: NASA/JPL-Caltech; Optical: NASA/ESA/S. Beckwith, Hubble Heritage Team (STScI/AURA); IR: NASA/JPL-Caltech/ University of Arizona/R. Kennicutt.

Day, 1925, a paper by Hubble (who strangely did not attend, but read in his absence), announced his discovery of Cepheids in spiral galaxies. The paper (Hubble 1925) derived a distance of 285 kpc for both M 31 and M 33, whilst modern day estimates are 780 kpc and 860 kpc, respectively. It was correctly announced at the time that spiral nebulae were extragalactic and were not components of our Galaxy. The true "Kantian" nature of the universe had been proven with the help of a special variable star.

1.3 Multiwavelength laboratories in space

Galaxies can be regarded as laboratories in space that provide information on widely differing phenomena such as star formation, the cold (T < 25 K),[10] warm (25 K < T < 20,000 K) and hot (T > 20,000 K) interstellar medium (ISM), stellar populations, high-velocity particles, non-thermal (non-stellar) activity and dark matter (more on dark matter in Section 2.5).

While observations of our own Galaxy allow the highest angular resolution inspection of nearby phenomena in a relatively normal spiral galaxy, our viewpoint is naturally restricted when observing other galaxies.

Galaxies vary in morphology, luminosity, mass (both luminous and dark), age, non-thermal activity, kinematics and star-formation properties. They also exist in widely varying environments that range from low-density regions to medium-density regions of groups of galaxies (with 3–50 members) to the crowded centers of rich clusters (numbering well over 1000 total members) with central densities of $\sim$100 galaxies per Mpc^2.

Therefore in order to accurately understand the physical properties of galaxies it is necessary to study a large variety of them. These varied properties should be studied in regions other than the traditional optical region of the electromagnetic spectrum[11] to achieve a complete multiwavelength picture (e.g. Figure 1.5 and compare to Figure 1.3):

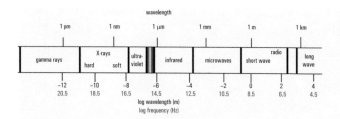

Figure 1.6

The electromagnetic spectrum. Prefixes are p – pico, 10^{-12}; n – nano, 10^{-9}; μ – micro, 10^{-6}; m – milli, 10^{-3}; and k – kilo, 10^3.

Credit: B. Lynds.

> They [galaxies] are to astronomy what atoms are to physics. ALLAN SANDAGE

Since the 1930s astronomers have been able to accurately detect radiation from galaxies in non-optical regions of the electromagnetic spectrum (Figure 1.6). To be precise, the first non-optical astronomical detection occurred much earlier than the 1930s. In 1800 William Herschel detected infrared radiation from the Sun. Now, more than 200 years later, all major regions of the electromagnetic spectrum – gamma ray, X-ray, ultraviolet, optical, infrared, submillimeter and radio – are being used to study galaxies (Figure 1.7):

> Trying to understand the universe through visible light alone is like listening to a Beethoven symphony and hearing only the cellos. JAMES B. KALER

This expansion into new regions of the electromagnetic spectrum has provided important discoveries about physical processes in galaxies, and greatly improved our understanding of previously known processes. In many cases the sensitivity and angular resolution of these multiwavelength observations can provide excellent images that allow detailed studies on relatively small spatial scales (e.g. 50–100 pc) in nearby galaxies (Figure 1.8). To put this spatial scale into perspective, many giant molecular clouds in nearby spiral galaxies that are stellar nurseries span ~100 pc. Imaging with this level of spatial resolution allows substructure such as nuclei, spiral arms, jets, tidal tails and rings of nearby galaxies to be investigated.

Figure 1.7

The radio galaxy NGC 5128/Centaurus A. Emission in X-rays is shown in blue, submillimeter in orange, and optical in white, brown.

Credit: X-ray: NASA/CXC/CfA/R. Kraft et al.; Submillimeter: MPIfR/ESO/APEX/A. Weiss et al.; Optical: ESO/WFI.

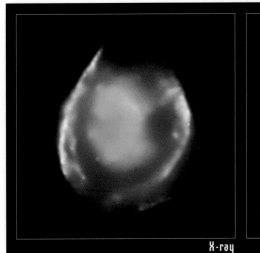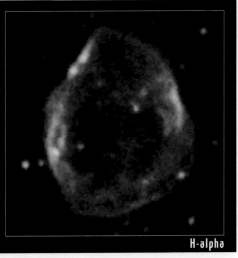

X-ray   H-alpha

Figure 1.8

X-ray and optical (Hα) images of the supernova remnant DEM L71 in the Large Magellanic Cloud. The side of the image is ∼50 pc at the distance of the LMC.

Credit: X-ray: NASA/CXC/Rutgers/ J. Hughes et al.; Optical: Rutgers Fabry-Perot.

1.4 The atlas galaxy sample

The choice of galaxies to include in this atlas was influenced by several factors. Firstly, the images had to be readily available from astronomers or from data archives. Secondly, such an atlas should aim to present images of galaxies that have good signal-to-noise and subtend large regions of the sky. Hence, many nearby galaxies (with distances less than 20 Mpc) that have been the target of extensive multiwavelength observations are included. Of the Local Group (the group of more than 35 galaxies within ∼1 Mpc of the Galaxy), the Andromeda Galaxy NGC 224/M 31, the Small Magellanic Cloud (SMC), NGC 598/M 33 (Figure 1.9), the Large Magellanic Cloud (LMC), NGC 6822 and of course the Galaxy are included.

When galaxies display interesting multiwavelength properties suggestive of unique astrophysical processes, or are representative of a certain class of galaxies, endeavors have been made to include them. This means including galaxies at distances greater than 20 Mpc even though they may have less extensive multiwavelength coverage or their images may have lower signal-to-noise.

1.5 Atlas galaxy categories

Our atlas galaxies are grouped into several, not necessarily mutually exclusive, categories. These are normal (N), interacting (I), merging (M), starburst (S) and active (A). The reasons for inclusion of a galaxy into a specific category is explained in the individual galaxy summaries in Part 4. Active galaxies typically have strong emission over a large portion of the electromagnetic spectrum making them prime targets for multiwavelength observations and thus they make up a large fraction of our sample.

1.5.1 Normal galaxies

Normal galaxies include galaxies that appear morphologically normal, do not possess unusual star-formation rates, and have continuum[12] spectra with a thermal (stellar) form characterized by one or more temperatures. However, upon close examination, few galaxies are completely "normal". For example, galaxies are seldom isolated from other galaxies and will often display some morphological signature of a dynamical disturbance. As well, many nearby galaxies that appear to be morphologically normal on the large scale are increasingly found to contain some type of low-luminosity active nucleus (e.g. the Galaxy, page 55; NGC 3031/M 81, page 177).

1.5.2 Interacting galaxies

Interacting galaxies display morphological signatures of a gravitational interaction with another nearby galaxy or are influenced by the passage through a dense medium that can "strip out" constituent gas. Such "ram pressure" stripping occurs frequently in cluster galaxies that are moving through a hot intracluster medium (ICM).

---

12 The general continuous shape of an object's spectrum, not including discrete lines.

Figure 1.9

NGC 598/M 33. An optical image made from three filters. Blue light is radiation through a B filter (~450 nm), green light through a V filter (~550 nm) and red light is from the hydrogen emission line Hα at 656 nm. The galaxy was observed with the KPNO 4 m and Mosaicl camera by Phil Massey (Lowell Observatory) and Shadrian Holmes (University of Texas).

Credit: P. Massey and the Local Group Survey Team.

Figure 1.10

NGC 4038/9: The "Antennae". An optical, wide-field view showing two gravitationally induced tidal tails.

Credit: B. Whitmore (STScI) and NASA.

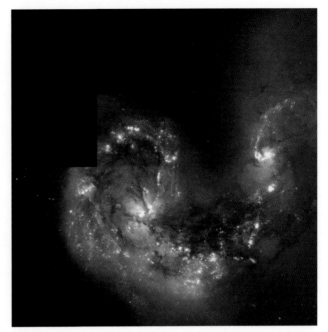

Figure 1.11

NGC 4038/9: An optical image showing the two colliding disks of the galaxies (in orange), criss-crossed by dark filaments of dust. A wide band of dust stretches between the centers of the galaxies. The spiral-like patterns, traced by bright blue star clusters, are a result of intense star-birth activity triggered by the collision.

Credit: B. Whitmore (STScI) and NASA.

Gravitationally induced morphological signatures can range from minor (e.g. spiral disk warps, isophotal[13] twists at large radii) to major (e.g. long, "tidal" tails of stars drawn out of a galaxy due to a close encounter). Disturbed, asymmetric gas distributions can suggest ram pressure stripping has occurred.

In the atlas the morphological changes vary from minor – NGC 4406/M 86 (page 119) in the Virgo Cluster displaying evidence of stripping of its ISM by the ICM – through to substantial – the dynamically interacting system of NGC 5194/M 51 and NGC 5195 (page 127). The Galaxy possesses a warp in its disk (Kerr 1957) which could be due to distortions in its dark matter halo by the orbital characteristics of the nearby Magellanic Clouds (Weinberg and Blitz 2006).

1.5.3 Merging galaxies

Merging galaxies are the later evolutionary stages of two or more interacting galaxies that have orbits and dynamics conducive to a final merger. Signatures that can suggest a merger include gravitationally produced tidal tails, disturbed morphologies, dust and gas when none should exist, and multiple nuclei.

The disk–disk merger NGC 4038/9 (page 146) seen in a wide-field image in Figure 1.10 shows long tidal stellar tails and two overlapping disks and Figure 1.11 shows the inner region remnants of the merging disks. Interestingly, direct collisions between stars in a merger are very rare, yet collisions between gaseous components occur and are very important. Our own Galaxy and NGC 224/M 31 are approaching each other and in several Gyr they will also undergo a major merger event.

1.5.4 Starburst galaxies

Starburst galaxies undergo intense star formation that is well in excess of normal rates. Large numbers of young stars (O and B spectral type) exist and the dust content can be extremely high. Normal star-formation rates

---

13 Isophotes are contours of equal light intensity.

(SFR) vary from essentially zero in elliptical galaxies to 10–20 $M_\odot$ yr^{-1} in some giant Sc galaxies (Kennicutt 1983). Total starburst SFRs of nearby galaxies can be 100 $M_\odot$ yr^{-1} and greater. These rates are similar to the highest peak SFRs predicted for protogalaxies (Kennicutt 1993) in which slow, fragmented collapse occurs (SFR$_{peak}$ ~10–100 $M_\odot$ yr^{-1}) and are lower limits to peak rates in which singular, rapid collapse (SFR$_{peak}$ ~100–1000 $M_\odot$ yr^{-1}) takes place. Thus the study of nearby starburst galaxies gives important clues to the early formation (protogalaxy) phases of normal galaxies. NGC 253 (page 154) and NGC 3034/M 82 (page 161) are prototypes of the starburst class. The case of NGC 3034/M 82 illustrates a possible link between two of our classifications, since its extreme SFR may be related to a recent (in astronomical terms) interaction with NGC 3031/M 81 (page 177). Figure 4.120 shows dramatic evidence of tidal interactions (filamentary features) between NGC 3031/M 81 and NGC 3034/M 82 (the dominant members of the M 81 Group) and NGC 3077.

1.5.5 Active galaxies

> All galactic nuclei are active, but some are more active than others. BERNARD PAGEL

Active galaxies have active galactic nuclei (AGN) and include LINER, Seyfert, radio galaxy, quasar and blazar types. "Activity" refers to non-stellar processes occurring or originating in a galaxy nucleus. An active nucleus has a spectrum with a continuum that cannot be explained by radiation from one or more stellar or blackbody[14] objects.

This non-thermal emission may be either synchrotron (produced by fast-moving charged particles in magnetic fields) or inverse Compton emission where a photon[15] collides with an energetic electron and some of the electron's energy is transferred to the photon. Likewise the Compton effect or scattering occurs when some of the energy of an X-ray or gamma-ray photon is transferred to an electron, leaving a less energetic photon.

In general, the spectra of active galaxies can be characterized by a power law

$$F_\nu = C\,\nu^{-\alpha}$$

where α is the power-law index, C is a constant, and F is the specific flux[16] (i.e. per unit frequency interval), usually measured in units of ergs s^{-1} cm^{-2} Hz^{-1}.

The ratio of the luminosity of the AGN to that of the whole galaxy can vary dramatically depending on the AGN type. Many normal looking galaxies may harbor mildly active nuclei in which the luminosity related to the activity may be a small percentage (~1% or less) of the total galaxy luminosity (e.g. LINERs). Quasars and blazars, on the other hand, have nuclear luminosities that can be more than several orders of magnitude larger than the luminosity of the *entire* host galaxy. Radio galaxy and Seyfert nuclear luminosities usually lie between these extremes.

Optical emission[17] lines from gas clouds near the nucleus are usually seen in AGN, are sometimes very strong and can have very large widths. The broadening of the lines indicates high velocities of the emitting gas clouds. Such galaxy nuclei must harbor some type of special energy source or central engine capable of vast energy production. Many AGN are believed to harbor supermassive black holes (SMBHs) with masses between a million and several billion $M_\odot$ (the latter in the cases of the most luminous elliptical galaxies).

Radiation from such active nuclei can also be greatly variable on time-scales as short as days or hours. This time variability can be used to estimate the size of the active region as further discussed later in this section. The main subclasses of active galaxies are described below.

LINERs

A significant fraction of galaxies exhibit low-level, nuclear activity and are classified as low ionization nuclear emission line regions (LINERs; Heckman 1980). Such galaxies, of both elliptical and spiral types, have nuclear optical spectra that cannot be reconciled with purely stellar emission, and are characterized by strong,

---

14 Blackbody radiation is determined only by the temperature of the object. Stars closely behave like blackbodies.

15 A photon is a bundle or quantum of electromagnetic radiation. Photons possess both wave-like and particle-like behavior. A simple way to reconcile this supposed inconsistency is to think of photons as "wave-packets".

16 Flux (density) is the energy of radiation passing through a given area in a given time.

17 A narrow region of the electromagnetic spectrum in which energy is emitted above the level of the adjacent continuum. Emission lines occur when energy is released during transitions between different energy levels in atoms or molecules. Similarly, absorption lines are gaps in the continuum and these occur because energy at particular wavelengths can be absorbed by atoms or molecules. The temperature, density and dynamical state of the emitters and absorbers can be inferred from the strengths of the lines.

OPTICAL

X-RAY

Figure 1.12

Optical and X-ray images of the
Seyfert galaxy NGC 1365. The inset
shows X-ray emission from the
central region of this active galaxy.
Using a number of separate X-ray
images astronomers can measure the
size of the hot, gaseous disk that in
theory feeds a central, supermassive
black hole.

*Credit: X-ray: NASA/CXC/CfA/INAF/
G. Risaliti; Optical: ESO/VLT.*

low-ionization[18] (with only a few electrons removed from the atom), forbidden[19] lines. LINERs are probably the lowest luminosity example of the AGN phenomena. The clearest signature of AGN phenomena, compact non-stellar continuum emission, is difficult to detect in LINERs since stellar emission dominates the observable flux.

Seyfert galaxies

Seyfert galaxies were discovered in the 1940s (Seyfert 1943) and were conspicuous due to their bright, "star-like" nuclei. Optical spectra of their nuclei show numerous, broad emission lines, indicative of high-velocity gas.

Seyferts are classified into two main groups. Seyfert 1 (Sy 1) types have Balmer lines (hydrogen) in emission that are broader than their emission lines of ionized metals.[20] The widths (usually denoted by full width half maximum or FWHM[21]) of the Sy 1 Balmer lines are often well in excess of 1000 km s^{-1}. Seyfert 2 (Sy 2) types possess Balmer and ionized metal lines of similar widths, typically 500–1000 km s^{-1}.

Antonucci and Miller (1985) have suggested that all Seyferts may be just one type of galaxy. They propose that NGC 1068/M 77 (page 171), the prototype Sy 2 galaxy, is in fact a Sy 1 with its nucleus enshrouded in dust, shielding its true Sy 1 nature. NGC 1365 (Figure 1.1 and Figure 1.12) is a Sy 2 galaxy. As discussed later

18 Ionization is the process of removing electrons from atoms or molecules. The particles are then positively charged. The amount of (positive) charge is dependent on the number of (negatively charged) electrons removed.

19 Forbidden lines are so called because they only occur in a low density ISM environment and are difficult to reproduce in terrestrial laboratories

where gas densities are much higher. They are not absolutely forbidden, yet the name remains.

20 In astronomy all elements higher in atomic weight than helium are called "metals" for historical reasons.

21 FWHM is the width of the point response function (PRF) at half the maximum intensity.

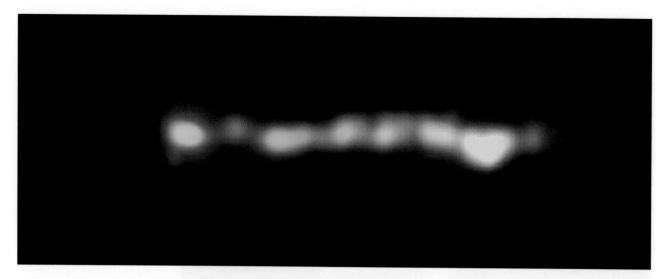

Figure 1.13

The 3C 273 jet. X-rays are shown in blue, optical in green, and infrared in red. Areas where optical and infrared light overlap appear as yellow. The center of the galaxy is to the left, outside of the image.

Credit: NASA/JPL-Caltech/Yale University.

in Section 2.8.1, it is very likely that all AGN are manifestations of the same general type of galaxy, as first suggested by the observational properties of Seyferts.

Radio galaxies

The prominent radio sources Centaurus A (also known as Cen A), Virgo A and Cygnus A were discovered from the late 1930s onwards, and reasonably accurate radio positions were determined by the late 1940s. Optical identifications followed from 1949 to 1954, which associated these sources with the host galaxies NGC 5128 (page 197), NGC 4486/M 87 (page 188) and a faint, elliptical galaxy in Cygnus (page 212), respectively. Such galaxies became known as radio galaxies.

Progress in radio observations was rapid and showed that some radio sources "straddled" their optical host galaxy with high-luminosity lobes of emission. These observations prompted theoretical studies on possible central energy sources (Blandford and Rees 1974) needed to power the large radio lobes. Increasingly sensitive observations coupled with higher angular resolution techniques (long-baseline interferometry – see Appendix A8) have provided fundamental information about radio galaxies. This includes the discovery of narrow, collimated jets of emission joining a central radio source to outer lobes at large radii.

One of these fundamental findings has been the differences in observed radio structures. Radio source morphologies show a dichotomy as a function of source luminosity (Fanaroff and Riley 1974), with the break luminosity occurring at $P^{\mathrm{tot}}_{1.4\mathrm{GHz}} \sim 10^{25}$ W Hz^{-1}. $P^{\mathrm{tot}}_{1.4\mathrm{GHz}}$ is the total power at the radio frequency[22] of 1.4 GHz, where hertz (Hz) is the unit of frequency, G is giga or 10^9 and watt (W) is a unit of power.

Radio sources of low power are generally "edge darkened" (Fanaroff–Riley Class I or FRI) whilst sources of high power are generally "edge brightened" (Fanaroff–Riley Class II or FRII). Centaurus A (page 197) is a FRI source whilst Cygnus A (page 212) is the archetypal FRII source.

A remarkable finding about compact nuclear radio sources was the discovery of superluminal motions (Moffet *et al.* 1972). Repeated observations of some sources (e.g. 3C 273, page 186) showed what appeared to be some source components separating at speeds well in excess of the speed of light, c. When the apparent separation or ejection velocity, v, exceeds c (defined as a transverse velocity $\beta_{\mathrm{app}} \equiv v/c > 1$), the motion is denoted superluminal. This apparent contradiction of physical law can be explained by a conspiracy between the orientation of the source motion and our line of sight to the object. Sources with a direction of ejection very close to our line of sight can possess true velocities as high as $0.99c$. Superluminal apparent motion of $9.6c$ has been detected in the inner part of the 3C 273 jet. The jet, observed by a variety of telescopes, is shown in Figure 1.13.

---

[22] Number of electromagnetic waves passing a fixed point in 1 second; symbol ν.

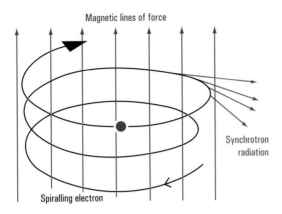

Figure 1.14

Synchrotron emission from fast-moving electrons spiraling in a magnetic field.

Radio galaxies emit radiation from a plasma[23] containing relativistic electrons. Relativistic in this case means that the electrons are travelling near to the speed of light, c. This radiation, known as synchrotron (Figure 1.14), is produced by decelerating electrons spiraling in strong magnetic fields. Strong radio sources also display synchrotron radiation in the optical region, as in the case of the NGC 4486/M 87 and 3C 273 jets.

Extended radio emission is seen as a variety of structures including emission on scales as large as several Mpc (galaxy cluster scale), large double lobes on opposite sides and outside the optical region of the galaxy (e.g. Cygnus A, page 212), highly collimated jets that sometimes end in broad lobes within the confines of the optical galaxy, single-sided jets, jets with brightness variations suggestive of sporadic particle ejection, precessing jets, and small-scale, $\sim$0.1 pc nuclear emission.

Quasars

Whilst many radio sources were identified with nearby galaxies in the early 1950s, some strong sources lacked optical identifications. A major breakthrough occurred in the early 1960s when the radio source 3C 273 (the host galaxy is shown in Figure 1.15) was positively identified with what appeared to be a faint, blue "star". Spectroscopic observations indicated an, at the time, extraordinary redshift[24] of 0.16 (Schmidt 1963) for this "star".

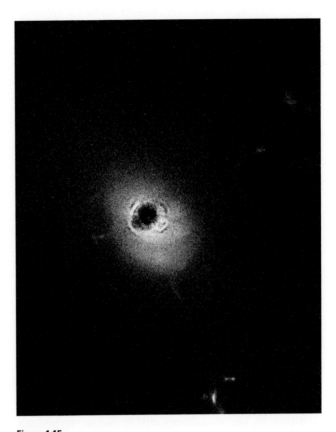

Figure 1.15

Optical image of 3C 273. The light from the luminous central quasar is blocked, allowing structures to be detected in the quasar's host galaxy, including a spiral plume and a red dust lane. The image is 30 arcseconds (or 85 kiloparsecs) across. North is $\sim$10° CW from up.

Credit: NASA, A. Martel (JHU), H. Ford (JHU), M. Clampin (STScI), G. Hartig (STScI), G. Illingworth (UCO/Lick Observatory), the ACS Science Team and ESA.

Such a high recession velocity implied a large, cosmological distance and thus this object was clearly not a nearby Galactic star. Spectroscopic observations of similar objects also found large redshifts. Adopting the large distances implied by the redshifts it was quickly realized that these relatively bright objects possessed the highest luminosities of any objects then observed in the Universe. For comparison, 3C 273 is about 1000 times more luminous in the optical region than our entire Galaxy.

Their star-like appearance led to the name of quasar, an amalgam of the term "QUASi-stellAR radio source". Paradoxically whilst radio observations allowed their discovery, radio-quiet (low-power) quasars are more numerous than their radio-strong (high-power) counterparts.

Optical studies also indicated variability in quasar emission. If the variability is caused by changes in the

---

23 A plasma is ionized gas in which electrons are not bound to atoms.

24 The change in wavelength of electromagnetic radiation due to the source moving with the expansion of the Universe. It is measured as the change in wavelength (observed minus emitted) divided by the wavelength of the radiation in the emitted rest-frame. It is denoted by the letter z.

nucleus structure itself, then the emission region will have a light-travel size similar to the time-scale of variability multiplied by c. Variability on time-scales as small as hours suggests such emission component sizes are similar to the size of our Solar System. For example, Pluto is ~5.5 light-hours from the Sun. Such small sizes add weight to the theory that the central engines are very compact. Optical spectra of quasars also show very broad (FWHM greater than 1000 km s^{-1}) emission lines, indicative of fast-moving gas clouds near the nucleus. Further dynamical measurements of stars and gas in the nuclear regions strongly suggested extremely large masses in these compact areas. The densities inferred were larger than any potential stellar aggregates. The majority viewpoint in astronomy is that these galaxies contain supermassive black holes with surrounding gaseous accretion disks.

Blazars

A variable blue "star" in the constellation Lacerta was discovered in 1941. Not until 1968, when it was shown to be a strong radio source, did it warrant significant attention. BL Lacerta (abbreviated to BL Lac) turned out to be quasar-like and not a star. Initial optical spectra were featureless and this hindered distance estimates. Other similar objects were found, and these displayed rapid variability (time-scales of days or weeks) and strong, time-varying polarization.[25] Large polarization values suggested a coherent, structural arrangement in the nucleus that influenced the preferred emission plane or direction of the radiation.

The generic term blazars was coined to describe such objects that display both low (BL Lac-like) and high (highly polarized quasars, HPQs, and the intriguingly named optically violent variables, OVVs) luminosities. The optical continuum emission is probably synchrotron emission that is intense because it is directed or beamed straight at us. Later spectra showed that optical emission lines from gas clouds are present in blazars but are very faint in comparison to the dominant continuum synchrotron emission. Very faint optical absorption lines from the galaxy hosts have also been detected and their redshifts show the hosts to be at large, quasar-like distances.

---

25 Light is polarized if photons have a preferred axis or plane of vibration of their electric field direction. Plane polarized light has an electric field vector in one plane only. Circularly polarized light has an electric field vector that rotates with constant angular frequency.

The all-sky view in gamma rays from the Fermi (formerly GLAST) satellite (Figure 1.16) shows a mixture of sources, both of Galactic and extragalactic origin. Two blazars, 3C 454.3 and PKS 1502+106, are detected. In December, 2009 Fermi detected a 10-fold brightening of 3C 454.3, making it three times brighter in gamma rays than the Vela Pulsar, and thus the brightest gamma-ray source in the sky. The blazar is also significantly brighter at visible and radio wavelengths and the changes in flux appear to be related to changes in the relativistic jet energetics and scattering of photons by relativistic electrons.

1.6 Properties of the atlas sample

Table 1 (page 18) lists the galaxies in the atlas. The table is divided into normal (N), interacting (I), merging (M), starburst (S) and active (A) categories. The table has column entries for galaxy name, group or cluster membership, right ascension, declination, morphological type, velocity, distance, other category and (atlas) page number. Apart from the entry for the Galaxy, all galaxies are ordered (in their respective category) by increasing right ascension.

1.6.1 Galaxy nomenclature

The galaxy names used in this atlas are taken from the New General Catalogue (NGC #; Dreyer 1888), Index Catalogue (IC #; Dreyer 1895, 1908), Messier Catalogue (M #; Messier 1784), Atlas of Peculiar Galaxies (Arp #; Arp 1966), the rich cluster Abell Catalogue (Abell #; Abell, Corwin and Olowin 1989), the radio survey Third Cambridge Catalogue (3C #; Edge *et al.* 1959; Bennett 1962), the Parkes radio source catalogue (PKS #; Ekers 1969) and from initial radio designations (e.g. Cen A; the brightest radio source in Centaurus). A cross-reference list is given in Table 5 (page 234).

1.6.2 Galaxy morphology

Galaxy morphologies can be broadly subdivided into elliptical (denoted by E), spiral (S), lenticular (S0) and irregular (Ir) classes. The first widely used morphology scheme was introduced by Edwin Hubble in 1926, which is now commonly known as the "Hubble sequence" or "tuning fork diagram". This initial scheme was updated with the inclusion of the S0 class in 1936 and Ir classes were later added as shown in Figure 1.17.

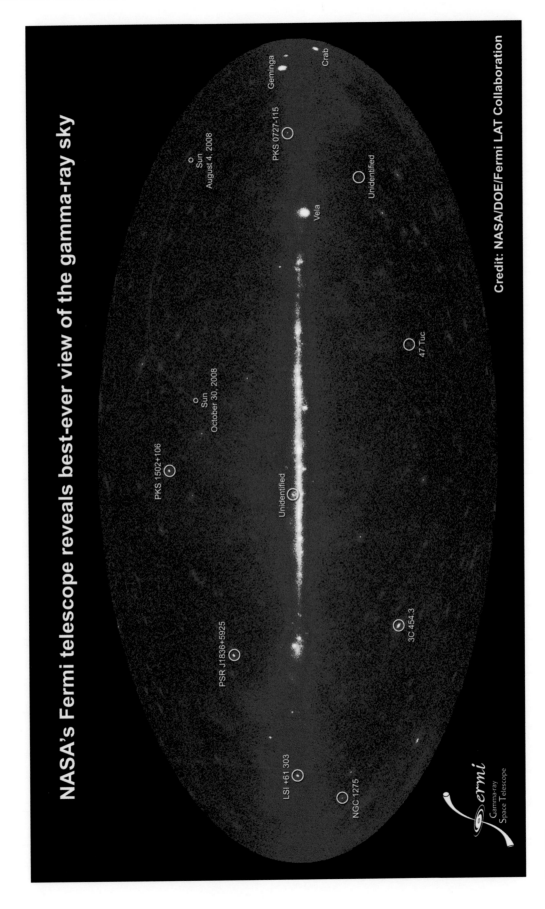

NASA's Fermi telescope reveals best-ever view of the gamma-ray sky

Credit: NASA/DOE/Fermi LAT Collaboration

Figure 1.16

Fermi all-sky view of gamma-ray sources. Galaxies include NGC 1275, blazars 3C 454.3, PKS 1502+106 and the quasar PKS 0727-115.

Credit: NASA/DOE/Fermi LAT Collaboration.

Table 1 The galaxy sample

Category/Galaxy	Group/Cluster	R.A. (1950) (2000)	Dec. (1950) (2000)	Type	v_0 (km s^{-1})	Dist. (Mpc)	Other Cat.	Page
Normal								
The Galaxy	LG	17 42 29 17 45 40	−28 59 18 −29 00 35	[Sbc/SBbc]		8.5E-3	I, A?	58
NGC 224/M 31	LG	00 40 00 00 42 44	40 59 42 41 16 07	SbI-II	−10	0.7		68
SMC	LG	00 50 53 00 52 38	−73 04 18 −72 48 00	ImIV-V	−19	0.07	I	75
NGC 300	Sth. Polar Grp.	00 52 31 00 54 53	−37 57 24 −37 41 09	ScII.8	128	1.2		78
NGC 598/M 33	LG	01 31 03 01 33 52	30 23 54 30 39 15	Sc(s)II-III	69	0.7		82
NGC 891	NGC 1023 Grp.	02 19 25 02 22 34	42 07 12 42 20 50	Sb on edge	779	9.6		90
NGC 1399	Fornax Cl.	03 36 35 03 38 30	−35 36 42 −35 26 58	E1	1375	16.9		94
LMC	LG	05 24 00 05 23 35	−69 48 −69 45	SBmIII	34	0.05	I	98
NGC 2915		09 26 31 09 26 12	−76 24 30 −76 37 36	[BCD]	468	3.3		105
Malin 2		10 37 10 10 39 53	21 06 29 20 50 49	[LSB]	13820	141		108
NGC 5457/M 101	M 101 Grp.	14 01 28 14 03 14	54 35 36 54 21 13	Sc(s)I	372	5.4	I?	110
NGC 6822	LG	19 42 07 19 44 57	−14 55 42 −14 48 24	ImIV-V	15	0.7		116
Interacting								
NGC 4406/M 86	Virgo Cl.	12 23 40 12 26 12	13 13 24 12 56 48	S0$_1$(3)/E3	−367	16.8		119
NGC 4472/M 49	Virgo Cl.	12 27 17 12 29 50	08 16 42 08 00 08	E1/S0$_1$(1)	822	16.8		122
NGC 4676		12 43 44 12 46 10	31 00 12 30 43 48		[0.022]	87.0		124
NGC 5194/M 51		13 27 46 13 29 52	47 27 18 47 11 50	Sbc(s)I-II	541	7.7	A	127
Merging								
NGC 520		01 22 00 01 24 35	03 31 54 03 47 31	Amorphous	2350	27.8		135
NGC 1275	Pers. Cl. Abell 426	03 16 30 03 19 49	41 19 48 41 30 38	E pec	5433	72.4	A	138
NGC 1316	Fornax Cl.	03 20 47 03 22 42	−37 23 06 −37 12 28	Sa pec (merger?)	1713	16.9	A	141
NGC 4038/9		11 59 19 12 01 53	−18 35 06 −18 51 48	Sc pec	1391	25.5		146

Table 1 (*cont.*)

Category/Galaxy	Group/Cluster	R.A. (1950) (2000)	Dec. (1950) (2000)	Type	v_0 (km s^{-1})	Dist. (Mpc)	Other Cat.	Page
NGC 7252		22 17 58 22 20 45	−24 55 54 −24 40 47	merger	4759	63.5		151

Starburst

Category/Galaxy	Group/Cluster	R.A.	Dec.	Type	v_0	Dist.	Other Cat.	Page
NGC 253	Sth. Polar Grp.	00 45 08 00 47 35	−25 33 42 −25 17 20	Sc(s)	293	3.0		154
NGC 3034/M 82	M 81 Grp.	09 51 41 09 55 50	69 54 54 69 40 40	Amorphous	409	5.2	I	161
NGC 5236/M 83		13 34 10 13 36 59	−29 36 48 −29 52 04	SBc(s)II	275	4.7	I?	165

Active

Category/Galaxy	Group/Cluster	R.A.	Dec.	Type	v_0	Dist.	Other Cat.	Page
NGC 1068/M 77		02 40 07 02 42 41	−00 13 30 −00 00 46	Sb(rs)II	1234	14.4		171
NGC 1365	Fornax Cl.	03 31 42 03 33 37	−36 18 18 −36 08 17	SBb(s)I	1562	16.9		174
NGC 3031/M 81	M 81 Grp.	09 51 30 09 55 36	69 18 18 69 04 04	Sb(r)I-II	124	1.4	I	177
NGC 4258/M 106		12 16 29 12 18 57	47 35 00 47 18 21	Sb(s)II	520	6.8		182
3C 273		12 26 33 12 29 06	02 19 43 02 03 08		[0.16]	589.0		186
NGC 4486/M 87	Virgo Cl.	12 28 17 12 30 49	12 40 06 12 23 32	E0	1136	16.8		188
NGC 4594/M 104		12 37 23 12 39 59	−11 21 00 −11 37 28	Sa$^+$/Sb$^-$	873	20.0		193
NGC 5128	N5128 Grp.	13 22 32 13 25 28	−42 45 30 −43 01 06	S0+S pec	251	4.9	I	197
A1795 #1	Abell 1795	13 46 34 13 48 52	26 50 25 26 35 31		[0.063]	243.9	I	206
Arp 220		15 32 47 15 34 57	23 40 08 23 30 11		[0.018]	71.3	S, M	209
Cygnus A		19 57 44 19 59 28	40 35 46 40 44 02		[0.057]	221.4		212

NOTES TO TABLE 1

- **Group/cluster:** LG: Local Group; Sth. Polar Grp.: South Polar Group, also known as the Sculptor Group; Pers. Cl.: Perseus Cluster.

- **Right ascension and declination:** From *A revised Shapley–Ames catalog of bright galaxies* (RSA; Sandage and Tammann 1981), when available, or else from the SIMBAD astronomical database except for the Galaxy (coordinates of Sagittarius A*, from Mezger *et al.* 1996).

- **Type:** Where possible galaxy types are taken from RSA. This classification scheme is the revised Hubble system (Sandage 1961, 1975). The exceptions, NGC 2915 (blue compact dwarf, BCD, Meurer, Mackie and Carignan 1994); the Galaxy (Sbc/SBbc, de Vaucouleurs 1970); Malin 2 (low surface brightness, LSB, McGaugh and Bothun 1994) are indicated with types given in [ ].

- **v_0:** The corrected recession velocity relative to the Local Group centroid, in km s^{-1} from RSA. Redshifts (z) are given in [ ] when v_0 is not listed in RSA. Velocities for NGC 2915 and Malin 2 are from the NASA/IPAC Extragalactic Database (NED) and are heliocentric.

- **Distance:** In Mpc from Tully (1988), except for the Galaxy (Mezger *et al.* 1996); LMC (50 kpc; note: Panagia *et al.* 1991; distance to SN 1987A, D(1987A) = 51.2±3.1 kpc); SMC (Rowan-Robinson 1985); Malin 2 (McGaugh and Bothun 1994); NGC 1275 and NGC 7252 (where $D = v_0/75.0$ Mpc); and NGC 4676, 3C 273, A1795 #1, Arp 220, Cygnus A where $D = v/75.0$ Mpc, and $v = c\frac{[(z+1)^2-1]}{[(z+1)^2+1]}$.

- **Other categories:** N – normal, I – interacting, M – merging, S – starburst, A – active. Explanatory notes on category classifications exist in Sections 1.5 and 1.6.4 and in Part 4 for each galaxy in the atlas.

- **Page:** Starting page number of the galaxy in the atlas.

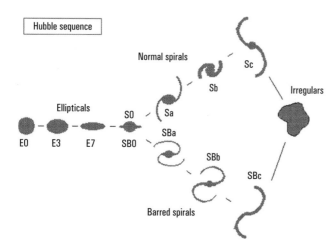

Hubble sequence

Normal spirals

Sc

Sb

Irregulars

Ellipticals

Sa

S0

SBa

E0   E3   E7   SB0

SBb

SBc

Barred spirals

Figure 1.17

The Hubble sequence of galaxy morphology. *Credit: J. Schombert.*

Elliptical and lenticular galaxies are sometimes referred to as "early-type", whilst spirals and irregulars are "late-type". These terms were used initially in the Hubble sequence to describe a probable evolutionary sequence from spirals with tightly wound arms (Sa; early-type) to spirals with more open arms (Sc; late-type). It is now known that the Hubble sequence is not an evolutionary sequence. Some galaxies, however, can change their morphology. For example, disk–disk (i.e. spiral–spiral) mergers can produce single elliptical-like galaxies. Changes in gas content in spirals by dynamical processes or normal ageing processes may also alter morphological classifications by several subclasses:

> An idea of how, for example, an arm in an Sb galaxy can change appearance in less than a rotation period, ..., can be simulated in a coffee cup by watching the structures "evolve" (change, branch, disappear and appear again) as the rotation of the coffee in the cup proceeds. ALLAN SANDAGE

The "tuning fork" morphological description does not completely encompass all types of galaxies (van den Bergh 1998). The discovery of the Sculptor and Fornax dwarf spheroidal galaxies in the mid-1930s began a still ongoing debate about the relation of higher luminosity ellipticals and their lower luminosity dwarf counterparts (not included in the original classification scheme). Dwarf galaxies are classified as spheroidal, elliptical or irregular and the dwarf spheroidal class is the most common type of galaxy in the Universe. The very large, luminous elliptical "cD" galaxies with extensive low surface brightness halos, found usually at the spatial and dynamical centers of rich clusters, are also overlooked.

Figure 1.18

Hubble Ultra Deep Field. Full-field optical image at left and subsection at top-right showing the position of the IR emitting galaxy HUDF-JD2. Center-right: HST uses NICMOS to view HUDF-JD2 in the near-IR. Bottom-right: Spitzer IRAC camera shows 3.6–8.0 μm emission from the $z \sim 6.5$ galaxy.

Credit: NASA, ESA, B. Mobasher (STScI/ESA).

Distant Galaxy in the Hubble Ultra Deep Field

Spitzer Space Telescope • IRAC
Hubble Space Telescope • ACS • NICMOS

NASA, ESA / JPL-Caltech / B. Mobasher (STScI/ESA)

ssc2005-19b

Figure 1.19

The Magellanic Cloud-type ImIV-V galaxy NGC 6822. This is an optical image made from three filters. Blue light is radiation transmitted through a B filter (~450 nm), green light is transmitted through a V filter (~550 nm) and red light is radiation from the hydrogen emission line Hα at 656 nm. The image was taken by the CTIO 4 m with the Mosaic-2 camera by Knut Olsen and Chris Smith (CTIO).

Credit: P. Massey and the Local Group Survey Team.

Clearly the original Hubble classification concentrated on nearby galaxies in a relatively narrow luminosity range. It was also natural for the higher surface brightness galaxies to be selected in a scheme both for ease of classification and observation. The detection of low surface brightness galaxies (LSBs;[26] Impey and Bothun 1987) beginning in the late 1980s suggests a bias in the original schemes. A "line of sight" bias also occurs for ellipticals as they are classified by their apparent (not true) axial ratios (see below).

Galaxy morphology is a function of lookback time or epoch – this is explored in more detail in Section 2.6. Evidence that morphologies are related to the epoch of observation come from high-resolution images from HST, including the Medium Deep Survey (MDS) and the Hubble Deep Field (HDF). The MDS detects galaxies with a ~5 Gyr lookback time and finds little gross change in the morphologies from those at the present epoch. However, the fraction of disturbed, distorted and asymmetrical galaxies in the HDF, at a ~8 Gyr lookback time, compared to the MDS, is roughly double. There is little change in the fraction of ellipticals, suggesting their general formation at much earlier epochs. There is also a lack of "grand[27] design" spirals in the HDF, suggesting that most if not all spirals had not yet formed. The Hubble Ultra Deep Field (HUDF; Beckwith *et al.* 2006) showed few galaxies at $z > 4$ (~11.5 Gyr lookback time) looking like present epoch ellipticals or spirals. The HUDF results confirm the viewpoint that the majority of initial galaxy assembly occurred within 1–2 Gyr of the Big Bang. One galaxy in the HUDF, denoted HUDF-JD2 (Figure 1.18), has $z \sim 6.5$ which suggests a lookback time of ~12.8 Gyr.

For this atlas, types are taken from *A revised Shapley–Ames catalog of bright galaxies* (Sandage and Tammann

---

26 Typically those galaxies that have a central surface brightness fainter than 23 B mag arcsec^{-2}.

27 Grand design spirals have well-defined spiral arm structure.

1981; RSA) where listed, which uses a revised Hubble sequence system (Sandage 1961). In particular the revised scheme includes intermediate classes (e.g. S0/a, Scd), Sm and Im classes (the "m" indicate varieties of Magellanic Cloud-type systems, with Sm or Im dependent on the presence or absence of vague spiral arms, respectively) and a new class "Amorphous" that replaces the irregular class. The Magellanic Cloud-type ImIV-V galaxy NGC 6822 is shown in Figure 1.19. The amorphous-type galaxy NGC 520 is shown in Figure 1.20:

> Galaxies are like people. The better you get to know them the more peculiar they often seem to become.
>
> SIDNEY VAN DEN BERGH

Stellar bars are denoted by "B" as are "pec" (peculiar), "merger" and "edge on". Inner ring structures (r) and spirals that connect to a bar (s) are indicated. E and S0s are classified E# and S0(#) by their apparent flattening, with the numeral # equal to $10(1 - b/a)$ and a and b are the major and minor axis lengths, respectively. E0s have a circular appearance, whilst the most flattened ellipticals observed are E7 types. Lenticulars can range from S0(0) to very flat S0(10). Spiral and very late-type galaxies (e.g. Im) are also classified into luminosity classes that measure the presence of spiral arms, surface brightness and pattern coherence. The earliest class is designated "I", the latest class, "V". For example, Sc I galaxies have very long organized arms, and Sc Vs have very poorly organized spiral patterns.

The morphological range for the spirals in the atlas is reasonably extensive, ranging from Sa pec (merger?) (NGC 1316) and Sa$^+$/Sb$^-$ (NGC 4594/M 104) and SbI-II (NGC 224/M 31) to Sc(s) (e.g. NGC 5457/M 101). Barred spirals (NGC 1365 and NGC 5236/M 83), Im (SMC, NGC 6822) and amorphous (NGC 520, NGC 3034/M 82) types are included. Ellipticals are confined to the predominantly round E0 (e.g. NGC 4486/M 87) to E1 (e.g. NGC 1399) classes. Since observations of the Galaxy are made from within its disk, it is difficult to accurately type it. However the Galaxy is generally regarded as an Sbc/SBbc (de Vaucouleurs 1970) with many astronomers favoring the presence of a Galactic stellar bar.

1.6.3 Distances and luminosities

The Sun is located ~8.5 kpc (~28,000 light-years) away from the Galaxy nucleus in the Orion or Local spiral arm, and very close to the plane of the disk. For comparison, the other dominant galaxy in our Local Group, the Andromeda Galaxy NGC 224/M 31, is 0.7 Mpc distant

Figure 1.20

The amorphous-type galaxy NGC 520. This is an optical image made from four filters. A g′ filter (blue), r′ filter (green), i′ filter (yellow), and Hα (red) are used. The image is 5.6′ across.

Credit: Gemini Observatory/Association of Universities for Research in Astronomy/ K. Roth (Gemini)/T. Rector (University of Alaska).

Figure 1.21

NGC 1316 (Fornax A). The optical galaxy is shown in blue, the radio lobes at 1.4 GHz are shown in red.

Credit: Image courtesy of NRAO/AUI.

(2.3 million light-years), or about a factor of 100 more distant than our Galaxy nucleus. NGC 4486/M 87, the central elliptical in the Virgo Cluster, is at ~17 Mpc (55 million light-years), or about 25 times further away than NGC 224/M 31:

> Miss Henrietta S. Leavitt of the staff of the Harvard Observatory had the gift of seeing things and making useful records of her measures. She began by finding in the Magellanic Clouds the miracle variable stars that have subsequently turned out to be extremely significant both for the exploration of extragalactic space and for the measurement of star distances throughout our own Milky Way system.
>
> HARLOW SHAPLEY

The most distant galaxy in the atlas is 3C 273 (Figure 1.15) with a redshift of 0.16, which implies a distance of ~590 Mpc or ~1.9 billion light-years. This is more than 800 times further away than NGC 224/M 31. The most distant galaxies so far observed by the Hubble Space Telescope and large ground-based telescopes are ~12 billion light-years away.

The optical luminosity[28] range of the sample spans five orders of magnitude. The faintest galaxy is NGC 2915 (8.3×10^7 $L_{\odot B}$). NGC 1275 (1.3×10^{11} $L_{\odot B}$) and NGC 4594/M 104 (2.6×10^{11} $L_{\odot B}$) are the brightest elliptical and spiral, respectively, whilst the quasar 3C 273 is the most luminous object in the sample at 3.0×10^{12} $L_{\odot B}$.

1.6.4 Multiple classifications

There are galaxies that could be placed in two or more atlas categories. In this case, galaxies are listed in the primary category that best describes it, as judged by the author. Galaxies with a classification ending with "?" are probable category members.

For example, whilst NGC 1316 (Fornax A, Figure 1.21 and page 141) shows evidence of activity in its nucleus

---

28 $L_{\odot B}$ is the luminosity of the Sun observed using the (optical) B filter.

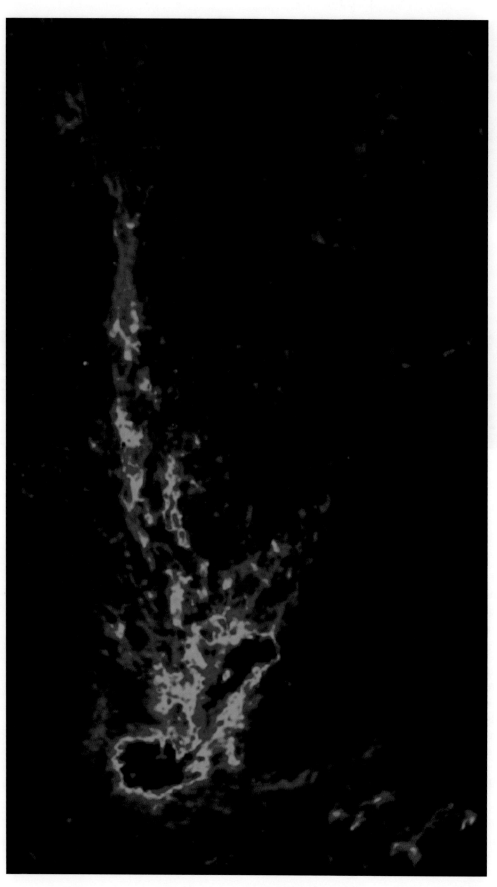

Figure 1.22

The radio H I Magellanic Stream starts at the bottom of the image where the large condensations indicate the positions of the LMC (left) and SMC (right) at declination $\sim -70°$ and extends upwards to declination $+02°$. Colors indicate increasing intensity of H I from red to blue. The image contains H I at velocities between -400 and $+400$ km s^{-1}.

Credit: M. Putman.

(LINER-type spectrum, nuclear radio jets), and possesses radio lobes at large radii, it shows dramatic evidence of recent mergers or interactions, based on the existence of low surface brightness shells, plumes, and extensive tidal tails as seen in Figure 4.98. NGC 1316 is regarded as a prototypical merger galaxy. In the atlas, NGC 1316 is listed as merging (M), yet it is also listed as an A (active; see Other category column in Table 1, page 18). Other examples of multiple classifications include the Large Magellanic Cloud (page 98), the Small Magellanic Cloud (page 75) and the Galaxy (page 58) which are all listed primarily as normal (N) yet have extra listings of I, due to the existence of the Magellanic Stream. This is a filament of neutral (atomic) hydrogen, H I (Figure 1.22) first discovered in the early 1970s (Wannier and Wrixon 1972) covering $100° \times 10°$ of the sky. It lies within the Galactic halo and contains $\sim 2 \times 10^8$ M$_\odot$ of H I (Putman *et al.* 2003).

The H I system extends from the LMC to the SMC via a connecting bridge, then extends more than $70°$ across the sky. Putman *et al.* (1998) discovered a leading arm feature (LAF) of gas moving in the opposite direction to the trailing feature of the Magellanic Stream. The most likely interpretation of the Magellanic Stream is that the LMC and SMC have passed close to, or through the halo of the Galaxy. Hydrogen gas has been gravitationally stripped out of the LMC and SMC and a stream is now attracted to the larger Galaxy. The Galaxy may also harbor a low-luminosity AGN (see Part 3), and thus warrants the additional classification of A? Multiple classifications are listed in Table 1 under Other Categories.

1.7 Epoch of formation and galaxy ages

Detecting a specific epoch of galaxy formation is still an open question in astronomy. For $\sim 300,000$ years after the Big Bang the universe was radiation dominated whereby a "soup" of energetic fundamental particles and radiation existed, but not atoms. This gives one limit to any epoch of formation. The most distant galaxy observed is 12.9 billion light-years away, thus giving it an age of at least 12.9 Gyr. This distant galaxy, IOK-1, has a firmly established redshift of $z \sim 6.964$ (Iye *et al.* 2006) and a star-formation rate of 10 M$_\odot$ yr^{-1}. Hence, if the Universe is 13.7 Gyr old (Dunkley *et al.* 2009) the first galaxies began to form between 300,000 and 800,000 years after

the Big Bang. Naturally, detections of more distant galaxies will further constrain the length of any "first period" of galaxy formation and therefore increase the probability of a special formation "epoch".

Relative galaxy-formation time-scales for different galaxies can be estimated by looking at the stellar-formation histories. Es and S0s are dominated by old stellar populations and little ongoing star formation is evident. It is generally thought that the bulk of their stars formed early in the age of the Universe. Spirals and irregulars, on the other hand, show clear evidence of ongoing star formation and mixed age stellar populations. A cartoon description of star-formation rates in ellipticals and spirals is shown in Figure 1.23.

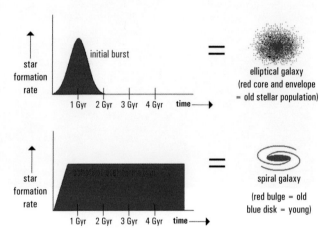

Figure 1.23

Star-formation rates for ellipticals and spirals.

Credit: J. Schombert.

Specifying a "single" or unique age of any given galaxy seems unphysical. Most galaxies have stellar subcomponents that have differing time-scales of formation. The Galaxy has been built via a hierarchical merger sequence and this merger history has constrained the star-formation history of various components (see Wyse 2009). It is thus more valid to discuss stellar age distributions of the subcomponents. The spheroid, created by major mergers and tidal debris, is old and was built between 8 and 12 Gyr ago. The central bulge is dominated by 10–12 Gyr old, metal-rich stars. The bulk of thick disk stars are 10–12 Gyr old, though it is likely built by ongoing satellite galaxy accretion into a pre-existing thin disk, which constrains the number and type of late accretion events. The thin disk appears to have a peak star-formation rate ~ 3 Gyr ago.

Figure 1.24

Optical image of the Large Magellanic Cloud. The image was constructed from more than 1500 separate images, using green and red continuum, and narrowband [O III] (blue), [S II] (green), Hα (red) filters. Star light (green and red continuum) is suppressed to enhance the interstellar medium of [O III], [S II] and Hα.

Credit: C. Smith, S. Points, the MCELS Team and NOAO/AURA/NSF.

Color–magnitude diagrams of resolved stellar populations combined with model isochrones (loci showing the position of stars of the same age) allow accurate age determinations. Spectroscopy of individual stars providing abundances of elements that radioactively decay, combined with stellar models, is another method to determine ages. This chronometric age gives us a lower limit to an age of the host galaxy. For example, several teams of astronomers have obtained high-quality spectral observations of a metal-poor halo star CS 22892−52 in our Galaxy. Sneden *et al.* (2003) derive an age of 12.8 ±3 Gyr from the Th/Eu abundance ratio. These ages were determined by measuring the abundances of uranium (upper limit) and thorium from absorption lines in the star and utilizing known decay times for these radioactive elements.

Appearances can be deceptive. It might be suspected that the type SBmIII LMC (Figure 1.24) is very young, based on its amorphous structure, large gas and dust content and numerous star-formation sites. However, age determinations (∼10–12 Gyr) of its oldest globular clusters are close to the ages of the oldest globular clusters in the Galaxy. Observations show the outer regions of the LMC are dominated by a ∼7 Gyr old stellar population. Star-formation rates also increased ∼4 Gyr ago, possibly due to the LMC collapsing from a spherical-like object to a more planar structure seen today. Ongoing star formation is supported by the cold atomic hydrogen gas structure (Figure 1.25). This image emphasizes the turbulent and fractal structure of the LMC ISM. The atomic ISM in the LMC is dominated by H I filaments, shells and voids, in many instances created by the injection of energy from recent supernovae and stellar associations. The total ages of both the Galaxy and the LMC may be similar, yet their individual star-formation histories differ.

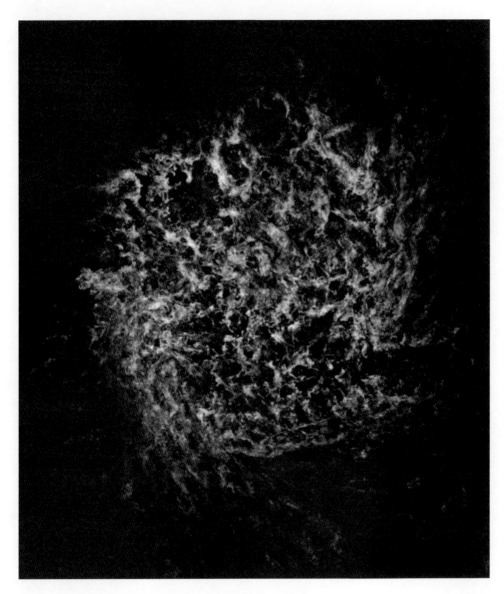

1.8 Additional reading

The listing is not intended to be comprehensive but will complement the text. Advanced texts or papers are indicated with the † symbol.

L. Belkora, 2003, *Minding the Heavens; The Story of our Discovery of the Milky Way* (Institute of Physics Publishing, Bristol).

K. Freeman and J. Bland-Hawthorn, 2002, The new Galaxy: Signatures of its formation, *Annual Review of Astronomy and Astrophysics*, **40**, 487.

S. van den Bergh, 1998, *Galaxy Morphology and Classification* (Cambridge University Press, Cambridge).

†J. Binney and M. Merrifield, 1998, *Galactic Astronomy* (Princeton University Press, Princeton).

†L. S. Sparke and J. S. Gallagher, III, 2007, *Galaxies in the Universe – An Introduction* (Cambridge University Press, Cambridge).

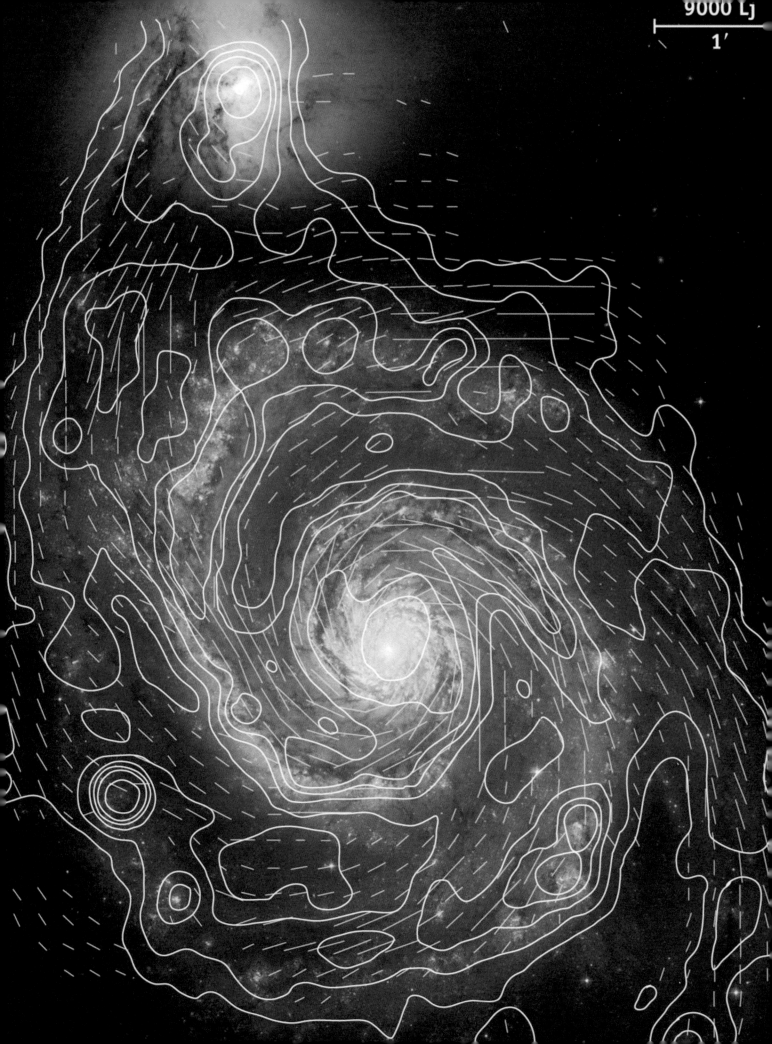

9000 Lj

1′

PART 2

Observing the electromagnetic spectrum

2.1 Earth's atmosphere and extraneous radiation

NOT all photons emitted by astronomical objects are detected by ground-based telescopes. A major barrier to the photons path is the Earth's ionosphere and upper atmosphere that absorbs or scatters most incoming radiation except for the optical (wavelengths of 3300–8000 Å where 1 Å $= 10^{-10}$ m), parts of the near-IR (0.8–7 μm) and radio (greater than 1 mm) regions. Absorption greatly affects radiation with the shortest wavelengths. In general, gamma rays are absorbed by atomic nuclei, X-rays by individual atoms and UV radiation by molecules. Incoming IR and submillimeter radiation are strongly absorbed by molecules in the upper atmosphere (e.g. H_2O and carbon monoxide, CO). Observations in these regions greatly benefit by locating telescopes at high altitude. Mountain-top sites like Mauna Kea (altitude 4200 m) in Hawaii, Cerro Pachon (2700 m), Las Campanas (2500 m) and Paranal (2600 m) in Chile, and La Palma (2300 m) in the Canary Islands are used to decrease the blocking effect of the atmosphere. The Antarctic, in particular the South Pole, provides an atmosphere with low water vapor content. Most of the continent is at high altitude, with the South Pole 2835 m above sea level, again helping to reduce the amount of obscuring atmosphere. The Antarctic has therefore also become a very useful IR and submillimeter site.

The transmission properties of the Earth's atmosphere (Figure 2.1) has prompted the exploration of the gamma ray, X-ray, UV, mid- and far-IR regions of the electromagnetic spectrum via satellite and high-altitude balloon observations. Satellite observations in the optical (e.g. Hubble Space Telescope) have benefited from

being above the majority of the atmosphere allowing near-diffraction limited observations. Radio astronomy satellites have benefited from large-distance instrumental baselines. HALCA (Highly Advanced Laboratory for Communications and Astronomy), known as Haruka after launch, operated from 1997 to 2003. It was an 8 m diameter radio telescope used for very long baseline interferometry. An elliptical orbit (21,400 by 560 km) allowed imaging by the satellite and ground-based telescopes, with good (u,v) plane coverage and high resolution. In late 1998 the Balloon Observations of Millimetric Extragalactic Radiation and Geophysics experiment (BOOMERANG), observed the sky at millimeter wavelengths for about 10 days. Future space interferometry missions include ESAs Darwin mission and NASAs Terrestrial Planet Finder, both in mission concept stages of planning.

Emission from the night sky plays an important part in observational astronomy and seriously affects our ability to detect faint objects. Reactions in the upper atmosphere that result in radiation are known as *airglow* or *nightglow*. Electrons recombining with ions (e.g. O, Na, O_2, OH) at typical altitudes of 100 km can radiate in the ultraviolet, optical and near-IR regions. The emission is usually measured in Rayleighs where

$$1 \text{ Rayleigh} = 10^6 \text{ photons cm}^{-2} \text{ s}^{-1} \text{ sr}^{-1}$$

and sr is steradian.[1] For example, at 762 nm the emission from O_2 is ~6000 Rayleigh.

The interaction of the solar wind with the Earth's magnetic field results in polar *aurorae*, usually close to

---

1 The steradian is a unit of solid angle. A sphere measures 4π ~12.56637 steradians.

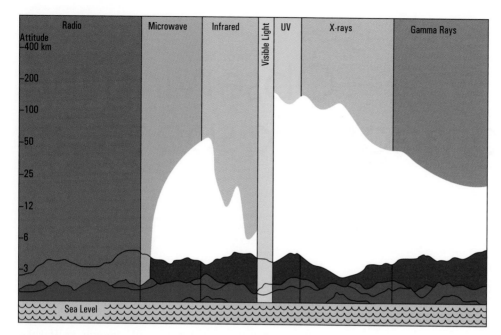

Figure 2.1

Transmission properties of Earth's atmosphere.

Credit: NASA and "Imagine the Universe!" http://imagine.gsfc.nasa.gov/.

the geomagnetic poles. Dust grains in the plane of the Solar System scatter sunlight causing *zodiacal light*. The Sun and the Moon are major contributors to night sky brightness. The influence of the Moon is easy to witness if you compare the night sky brightness at both Full and New Moon. Optical observations that aim to detect very faint objects are usually scheduled during dark skies, close to New Moon.

Other extraneous radiation sources are also present. Ground-based near-IR observations are plagued by background heat radiation from the telescope (e.g. mirrors) and structure (e.g. oil lubricated bearing of horseshoe mounts; Figure 2.2). The insert image of Figure 2.2 at a wavelength of 10 μm shows temperature changes represented by color differences. Oil lubricated bearings that support the horseshoe mount appear in the infrared image as bright red, representing a 15 °C increase in temperature above surrounding structures.

This nuisance radiation can be minimized by cooling and reducing the surface area along the optical path of the instruments. The near-IR background sky is also very bright and highly variable on short time-scales. Observations from the excellent ground-based IR sites of the Antarctic, Chile and Mauna Kea can help minimize such fluctuations, yet satellite observations offer the best IR observing conditions.

Human activity produces spurious radiation sources that can affect astronomical observations. These include microwave and radio emission from industrial and telecommunication sources. In the optical region night-time outdoor lighting and general city and suburban lights have all put additional pressure on the quality of observations. Finally, whilst satellites have allowed us to make observations across the entire electromagnetic region, they too are sources of increasing "pollution" for ground-based observations, when recorded as streaks of light across long-exposure wide-field images near to, or through (Figure 2.3) objects of interest. Of July, 2009 there were ~900 operational satellites, with ~1500 objects greater than 100 kg in mass in orbit, and 19,000 objects in orbit with diameters >10 cm.

2.2 Temperature, energy, wavelength and frequency

In astronomy, temperatures are usually quoted in terms of kelvin (K). One K is the same interval as 1 degree Celsius (°C), however the kelvin scale starts at absolute zero, or −273.16 °C.

As might be expected, temperatures of astronomical radiation sources vary widely. Dusty, dark nebulae (e.g. the Horsehead Nebula in Orion) exist at temperatures between 10 and 100 K enabling molecular hydrogen (H_2), carbon monoxide (CO), hydrogen cyanide (HCN) and water (H_2O) to exist in molecular clouds.

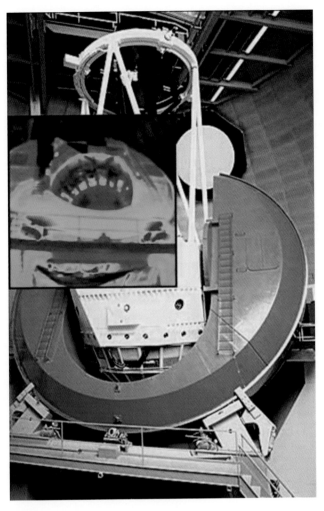

Figure 2.2

The main image shows the Mayall 4 m telescope at Kitt Peak National Observatory in visual light. The insert shows the horseshoe mount at 10 μm taken with a thermal video camera. Hot oil lubricated bearings that support the mount appear as bright red.

Credit: National Optical Astronomy Observatory, M. Hanna, G. Jacoby.

Dust grains emit at a characteristic temperature between 20 and 100 K and are found in and near such clouds. The temperature of neutral or atomic hydrogen, H I, is usually between 25 and 250 K. Emission nebulae or ionized H II regions (near hot, young stars that strongly emit UV radiation) exist at ~10,000 K. The surface temperatures of stars range from 2500 to 40,000 K. Our Sun, a G dwarf, has a surface temperature of 5800 K. Surface temperatures of neutron stars could be several $\times 10^5$ K. Gas temperatures in accretion disks (e.g. around the black hole candidate Cygnus X-1) are $\sim 2 \times 10^6$ K. Galaxy cluster (ICM) gas, detected in the X-ray region, typically has temperatures of 10^7 K. The temperature of gas involved in thermonuclear explosions near the

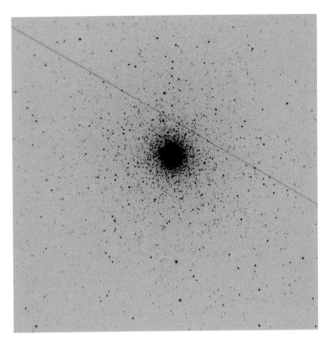

Figure 2.3

Globular cluster NGC 104 (47 Tuc) and satellite trail. A 30 second exposure, 30″ across.

Credit: A. Mattingly, Grove Creek Observatory.

surface of accreting neutron stars is $\sim 10^{7-9}$ K. Hence astronomical object temperatures range over nine orders of magnitude, or a factor of a billion.

Emission, especially in the gamma ray and X-ray regions, is typically measured in terms of its corresponding energy in units of electron volts[2] (eV). X-ray energies are usually measured in terms of keV with kT as the symbol for energy, where the k (in kT) is Boltzmann's constant $= 1.38 \times 10^{-16}$ erg K^{-1}. Gamma ray energies are quoted in MeV, GeV and in extreme cases TeV.

Radiation wavelengths are typically quoted when discussing the EUV region and longer wavelengths. Nanometers (nm) and angstroms (Å) are used until the near-IR when microns (μm) are stated. From submillimeter to radio, units progress from mm to cm to m and are interchanged with frequency units such as MHz and GHz.

Table 2 describes the key events in multiwavelength observations of galaxies beginning with Jansky's detection of radio emission from the Galaxy in 1931.

---

2 One eV is the energy acquired by an electron when it is accelerated through a potential difference of 1 volt in a vacuum. 1 eV has an associated energy $= 1.60 \times 10^{-12}$ erg.

Table 2 Key events in multiwavelength observations of galaxies

Year	Event
1931–1933	Jansky detects radio emission from the Galaxy
1939	Reber detects the radio source Cygnus A
1943	Seyfert identifies six spiral galaxies with broad emission lines
1949	Identification of radio sources Virgo A (with M 87) and Centaurus A (NGC 5128) by Bolton, Stanley and Slee
1951	Discovery of 21 cm emission from interstellar hydrogen by Ewen and Purcell
1954	Baade and Minkowski identify the optical counterpart of Cygnus A
1961	Explorer XI satellite detects gamma rays
1962–1966	Aerobee rockets detect X-ray sources (including M 87)
1963	Identification of quasars
1967	Gamma ray bursts (GRBs) detected by Vela satellites
late 1960s	First major IR survey by Neugebauer and Leighton detects $\sim$6000 near-IR sources
1968–1972	OAO series of satellites detect UV sources
1978–1980	HEAO-2 (Einstein) increases number of extragalactic X-ray sources
1983	IRAS performs sky survey at 12, 25, 60 and 100 μm
1985	Antonucci and Miller discover Sy 2 NGC 1068 has broad emission lines (similar to Sy 1s) in polarized light
1987	James Clerk Maxwell Telescope (JCMT) opens
1990	Launch of Hubble Space Telescope, ROSAT
1991	Launch of Compton Gamma Ray Observatory (re-entered in 2000)
1992	mm observations by COBE detects 30 μK deviations in the cosmic background radiation
1995	Hubble Deep Field-North observations, ISO launched
1997	Distance scale to GRBs determined via X-ray, gamma ray and optical observations
1999	Launch of Chandra X-ray Observatory, XMM-Newton
2001	2MASS Near-IR survey ends (began 1997)
2003	Wilkinson Microwave Anisotropy Probe (WMAP) First Data Release, Spitzer Space Telescope and GALEX launched
2004	Swift launched
2008	Fermi Gamma Ray Space Telescope launched
2009	Herschel Space Observatory, Wide-field Infrared Survey Explorer (WISE) launched

2.3 Astronomical sources of radiation

The entire electromagnetic spectrum (Figure 1.6) stretches more than 15 orders of magnitude (a factor of 10^{15}) from short wavelength ($\sim 10^{-10}$ cm) high-frequency gamma rays through X-rays, ultraviolet, optical, infrared, submillimeter to the longest wavelengths ($\sim 10^{5}$ cm) and lowest frequencies of the radio region.

The observational limits of electromagnetic radiation from astronomical objects are not well established. At high energy, short wavelengths, TeV (where T is tera or 10^{12}) gamma rays have been detected from some AGN. In comparison, the highest energy cosmic ray particles (typically protons), above 10^{19} eV (or 10^{7} TeV), arrive at a rate of about one particle per square kilometer per year. Low-to-medium energy cosmic rays, up to energies of about 10^{18} eV, probably originate in the Galaxy via interactions with magnetic fields. Higher energy cosmic rays are most likely extragalactic in origin, possibly in AGN or supernovae.

At low energy, long wavelengths, radio radiation has been detected at wavelengths of about 1.2 km (frequency 0.25 MHz; Novaco and Brown 1978). The long wavelength limit of a few km is set by absorption in the interplanetary and interstellar media.

What types of objects emit radiation and in what region of the electromagnetic spectrum is it detected? Table 3 (page 34) describes the main spectral regions and lists astronomical sources of emission and absorption in each region.

> You know what finally happened. ... I concluded that we had to distinguish at least two H-R diagrams – one the normal diagram that we had known well for some time, the other, the globular-cluster diagram.
>
> WALTER BAADE

2.4 Origin of astronomical radiation

In the following section the main sources of astronomical radiation will be described. (Telescopes and instruments are described in Appendix A.)

2.4.1 Gamma rays

$kT > 500$ keV

Observations of gamma rays are the most difficult of all multiwavelength detections and accurate identification of their origin is still debatable in some cases. The detection of faint extragalactic sources is difficult because the photons have to be detected against a high background of cosmic rays.[3] The angular resolution of gamma-ray telescopes is presently quite low, and optical identification of gamma-ray sources (especially when associated with faint optical sources) has proven very difficult.

However, Porter *et al.* (2009) show, for the first time, an external galaxy resolved in gamma rays. The Fermi LAT has resolved the gamma-ray emission from the LMC. The LMC is observed with an integration time of 211.7 days with energies between 200 MeV and 100 GeV and the gamma-ray signal is dominated by emission from the star-forming region 30 Doradus. The overall gamma-ray emission does not seem to correlate with the molecular gas distribution but better matches the atomic H I distribution.

Detections in other wavelengths can help pinpoint the origin of gamma-ray sources. For example, the Vela and Crab Pulsars emit pulsed radiation in both the gamma-ray and radio regions with the same periodicity, allowing certain gamma-ray source identification. Pulsars can produce gamma-ray emission if material falls onto their surface and is heated to temperatures of a few 10^{6} K.

Strong gamma-ray sources in our Galaxy include the Galactic plane, and several nearby pulsars (Figure 1.16 and Figure 3.5). The emission from the plane can be accounted for by inelastic collisions between high-energy cosmic rays, probably protons, and the nuclei of atoms and ions in interstellar gas. Such collisions result in the production of π mesons[4] which decay to two 70 MeV gamma rays.

This decay mechanism probably explains the high-energy (>100 MeV) events, whilst at lower energies Bremsstrahlung radiation (see below) could greatly contribute (see Longair 2010 for a detailed discussion).

The high X-ray luminosities of AGN strongly suggest that they should also be sources of gamma rays. This is now confirmed by observations. COMPTEL observations suggest the existence of "MeV quasars" that may contribute substantially to an MeV "bump" in

---

3 Energetic particles travelling close to the speed of light. Primary cosmic rays originate beyond the Earth's atmosphere. Secondary cosmic rays are produced when primary cosmic rays collide with atmospheric atomic nuclei, and are detected as air showers.

4 The strong nuclear force binds together protons and neutrons, involving the exchange of short-lived particles called mesons.

Table 3 Astronomical sources of emission and absorption

Spectral regions	Emission: Stellar	Emission: Interstellar	Absorbers
Gamma ray ($kT > 500$ keV)	Pulsars, bursts?	ISM scattered	
Hard X-ray (3 keV $< kT < 500$ keV)	X-ray binaries; AGN?	Ultra-hot ISM, ICM	
Soft X-ray ($0.1 < kT < 3$ keV)	Main seq. stars; evolved SNe	Hot ISM, ICM; SNRs	Dust
EUV (100–912 Å)	O Stars; Pop. II, evolved; accreting binary stars	Hot ISM, SNRs	Dust; H, He
Far-UV (912–2000 Å)	Pop. I, $M > 5\,M_\odot$; Pop. II, evolved	H II, Lyman alpha, Planetary nebulae	Dust, metals H_2, Lyman alpha
Mid-UV (2000–3300 Å)	Pop. I, $M > 1.5\,M_\odot$; Pop. II, horiz. branch	–	Dust, metals Ionized species
Optical (3300–8000 Å)	Pop. I evolved (BA, M); $M > 1\,M_\odot$ Main seq., Pop. II, Main seq.; Evolved (K,M): horiz. branch	H II, H-Balmer; Forbidden metal Emission lines, [O],[S],[N],[Ne]	Dust, metals
Near-IR (0.8–7 μm)	Evolved red giants, Supergiants, AGB stars; Protostars	H II; Hot dust, PAHs; H_2 emission	Dust; PAHs
Mid-IR (7–25 μm)	Hot circumstellar dust, OH/IR stars; protostars	H II; PAHs, Small grains	Dust
Far-IR (25–300 μm)	Carbon stars, protostars	H II; dust	Dust
Sub-mm (300 μm to 1 mm)	–	Dust; therm. brems. Molecular ISM; non-thermal	–
Radio (>1 mm)	–	Non-thermal, therm. brems.; Molecular ISM (CO, etc.) Masers Neutral ISM (H I) H II	–

NOTES TO TABLE 3

- This table is adapted from Table 1 of Gallagher and Fabbiano (1990), with kind permission of Springer Science and Business Media.

- Units/symbols: Gamma rays are typically described in units of T(era)eV, G(iga)eV, M(ega)eV. kT is a symbol for energy, typically used in X-ray astronomy, where k (Boltzmann's constant) $= 1.38 \times 10^{-16}$ erg/K; X-rays in k(ilo)eV; EUV, far-UV, mid-UV and optical in Å or nm; near-IR, mid-IR, far-IR in μm; sub-mm in μm, mm and GHz; radio in cm, m, MHz and GHz.

- Whilst CO observations are usually regarded as "mm" observations, to be consistent with the above spectral region divisions, CO observations at wavelengths longer than 1 mm will be considered as radio observations.

- Pop. I – Population I are stars and clusters that are relatively young, and are typically found in the Galactic disk region. They are formed from enriched material from previous stellar generations and they tend to have high metallicities.[5]

- Pop. II – Population II are stars and globular clusters that are relatively old, and are typically found in the spherical halo of the Galaxy. They have low metallicities, being formed earlier than Pop. I from less enriched material.

- X-ray binaries – Binary star systems in which one component is a degenerate star (e.g. white dwarf, neutron star or black hole). X-rays are emitted from either a gaseous accretion disk in low-mass X-ray binaries (LMXBs; when the two stars are of similar masses an accretion disk forms around the degenerate star) or from an extended envelope in high-mass X-ray binaries (HMXBs; when one component is $\sim$10–20 $M_\odot$ and gas flows directly onto the degenerate component).

- PAHs – Polycyclic aromatic hydrocarbons, a component of interstellar dust made up of small dust grains (e.g. silicates) and soot-like material. "Hydrocarbon" refers to a composition of C and H atoms. "Polycyclic" indicates the molecules have multiple loops of C atoms. "Aromatic" refers to the kinds of bonds that exist between the C atoms. PAHs are formed during incomplete combustion of organic (i.e. carbon-based) material. Observations of the "Red Rectangle" (HD 44179) nebula by Vijh, Witt, and Gordon (2004) showed blue luminescence at $\lambda < 5000$ Å. The authors attribute this to fluorescence by PAH molecules with three to four aromatic rings such as anthracene ($C_{14}H_{10}$) and pyrene ($C_{16}H_{10}$). However, Nayfeh, Habbal and Rao (2005) also suggest that ultrasmall silicon nanoparticles of 1 nm in diameter could be the source of emission.

- SNRs – A supernova remnant (SNR) is the remains of a supernova explosion. Massive stars end their lives by imploding, and the outer layers of gas are blown outwards at velocities up to 15,000 km s^{-1}.

---

5 The metallicity or metal abundance measures the amount of elements other than hydrogen or helium in a star or gas. This is typically expressed for a star as relative to the Sun as $[Fe/H] = \log_{10}\left(N_{Fe}/N_H\right)_{star} - \log_{10}\left(N_{Fe}/N_H\right)_\odot$ where N_{Fe} and N_H are the number of iron and hydrogen atoms per unit volume respectively.

the gamma-ray background[6] spectrum. EGRET observations have discovered bright, variable gamma-ray emission from blazars. These sources are identified with highly polarized OVVs (Optically Violent Variables) in which relativistic beaming is most likely occurring. Even higher energy emission has been detected from some AGN. Gamma rays exceeding 5 ±1.5 TeV have been detected from Markarian[7] 421 (a BL Lac object; Krennrich *et al.* 1997) using the Whipple Observatory's 10 m telescope in Arizona.

The gamma-ray ultra-high energy emission is related to a jet or beamed energy originating in a putative SMBH and surrounding accretion disk. Gamma rays are produced by the beam of relativistic particles which is ejected and collimated by strong magnetic fields in the inner accretion disk region. The highest energy emission is seen in such objects when the jets are viewed end-on, which is occurring in blazars like Markarian 421.

Whilst gamma rays are absorbed by our atmosphere, very-high-energy (VHE) gamma rays can be detected from the ground via the secondary radiation they produce when they strike components of the Earth's atmosphere. This radiation is produced as a brief flash of light that only lasts for a few billionths of a second. This light can be detected with large optical light collectors equipped with photomultiplier tubes as on the Whipple Observatory 10 m telescope.

Gamma-ray astronomy is now heavily focussed on gamma ray bursts (GRBs; Fishman 1995; Paczyński 1995). These intense outbursts vary in duration from a few milliseconds to a few tens of seconds. GRBs are now known to be extragalactic in nature and appear to occur in the outskirts of distant galaxies. Gamma ray bursts are discussed in more detail in Section 2.8.4.

2.4.2 X-rays

Hard: 3 keV < kT < 500 keV; soft: 0.1< kT < 3 keV

The soft X-ray region (0.1< kT < 3 keV) can also be subdivided into smaller energy regions (e.g. see the 0.2–1.5 keV CXO observation of NGC 253 on page 156). X-rays originate from a variety of physical processes.

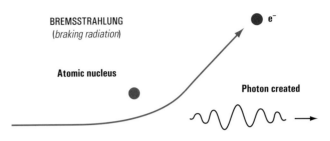

Figure 2.4

Bremsstrahlung radiation.

First, gas that is hotter than 10^6 K is fully ionized (i.e. all electrons are dissociated from their atoms). The electrons are accelerated due to proton encounters and radiate energy via the bremsstrahlung process. This radiation is emitted by hot gas associated with SNRs and by hot gas that surrounds many elliptical galaxies and galaxy clusters.

Bremsstrahlung, or braking radiation (Figure 2.4), occurs when charged particles, typically electrons, are decelerated over a very short distance. At temperatures higher than 10^5 K gas consists of positive ions and electrons. In bremsstrahlung, a continuous spectrum with a characteristic profile and energy cutoff (wavelength minimum) is produced:

$$I(E,T) = C\, G(E,T)\, Z^2\, n_e\, n_i (kT)^{-1/2}\, e^{-E/kT}$$

where $I(E,T)$ is the intensity (a function of energy, E, and temperature, T), C is a constant, G is the "Gaunt factor" (a slowly varying function), Z is the charge of the positive ion, n_e is the electron density, and n_i is the positive ion density.

This emission is characterized by the temperature of the gas. The higher the temperature, the faster the electrons, and the higher the photon energy of the radiation. Bremsstrahlung radiation is also known by astronomers as free–free emission, since the electron starts free and ends free.

Second, electrons spiraling in a magnetic field will emit synchrotron radiation. Synchrotron X-ray emission (Figure 1.14) requires very energetic, high-velocity electrons in strong magnetic fields. An example of this process is found in the SNR Crab Nebula. The intensity is of the form

$$I(E) = C\, E^{-\alpha}$$

where the intensity I is only a function of energy, E, C is a constant, and α is the spectral "index". Larger values

---

6 The gamma ray background is the integrated emission from sources in the gamma-ray region that are not resolved. These unresolved sources could be very faint or diffuse or both.

7 Markarian galaxies are galaxies catalogued by the astronomer B.E. Markarian based on their strong continuum emission in the UV.

of α correspond to a higher proportion of lower energies emitted.

Third, X-ray emission can be blackbody radiation. In this case, an object is called a "blackbody" if its surface re-emits all radiation that it absorbs. The continuum emission radiated is described by only one parameter, the objects temperature. X-ray blackbody radiation is emitted from very hot objects with surface temperatures $>10^6$ K, such as neutron stars. The intensity is given by the Planck law

$$I(E, T) = 2\,E^3\left[h^2c^2\big(e^{E/kT} - 1\big)\right]^{-1}$$

where the intensity is a function of energy, E, and temperature T, h is Planck's constant, and c is the speed of light.

Observations of the Galaxy and Local Group members suggest that much of the total X-ray emission from spiral galaxies originates from discrete sources such as accreting binaries and SNRs. Diffuse emission has been detected in many spirals and originates from hot gas energized by shocks or outflows (e.g. caused by supernovae) in their disks. Hot, 10^{6-7} K, gaseous halos around elliptical galaxies and clusters of galaxies were discovered by *Einstein*. These originate from accumulated ejected gas (mass loss) from the evolved stellar population, as confirmed by the enriched metal content of the gas. Characteristics of elliptical galaxy X-ray emission also suggest an underlying discrete source component, most likely from accreting binaries.

Figure 2.5 shows a CXO image of the central regions of the merging galaxies NGC 4038/9. The bright point-like sources are binary systems containing either neutron stars or black holes which are accreting gas from donor stars. The X-ray emission originates from accretion disks around these degenerate stars. Other more extended X-ray emission is associated with hot gas energized by numerous supernova explosions, stimulated by the merger process.

X-ray emission is also very strong in AGN and originates from a variety of sources. A component can be linked with frequently observed beamed radio and gamma-ray emission. X-ray emission with low (<1 keV) energies can be variable on time-scales as short as several hours. This emission probably originates close to the active core of the AGN, most likely in the inner region of an accretion disk. Higher energy X-ray emission in AGN is caused by inverse Compton scattering (Figure 2.6) which is an exchange of energy between

Figure 2.5

CXO X-ray image of the merging galaxies NGC 4038/9. The image is 4′ on a side.
Credit: NASA/SAO/CXC/G. Fabbiano.

electrons and photons in dense gas although for quasars this component is weak or absent.

An important emission line is seen in many AGN at 6.4 keV due to X-ray fluorescence[8] from iron at temperatures of several million K. Photons striking the accretion disk of the AGN are absorbed by electrons of iron which then de-excite by emitting a photon of energy 6.4 keV. This Fe Kα line is an important diagnostic of the kinematics of accretion disk gas near the central black hole. The ASCA satellite made a deep exposure of the Seyfert 1 galaxy MCG-6-30-15 (Tanaka *et al.* 1995) showing a

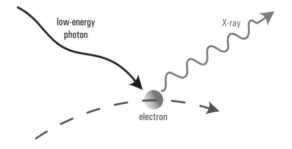

Figure 2.6

Inverse Compton scattering.

8 Certain substances can absorb radiation at one wavelength and re-emit it. Usually the fluorescent-based emission is at a larger wavelength and has less energy.

broad Fe Kα line that would imply relativistic gas speeds of $\sim$100,000 km s^{-1} or $0.3c$. The line profile is sometimes asymmetric, consistent with relativistic effects and provides compelling evidence for not only the existence of SMBHs but also black hole spin (Miller 2007).

2.4.3 Ultraviolet

Extreme (EUV): 100–912 Å; far-UV: 912–2000 Å; mid-UV: 2000–3300 Å

The EUV region is dominated by emission from (in order of decreasing number of detections in the second EUVE Source Catalog) late-type stars (F to M spectral classes), hot white dwarf stars, early-type stars (A, B spectral classes), cataclysmic[9] variables and AGN (mainly Seyferts and BL Lacs). EUV observations (i.e. ROSAT WFC and EUVE) consist of medium angular resolution surveys that generally detect objects in the Galaxy.

Based on the soft X-ray properties of some AGNs, many active galaxies were predicted to be detected in the ultraviolet. In fact WFC detected seven AGN (three were blazars) whilst Marshall, Fruscione and Carone (1995) detected 13 AGN (seven Seyferts, five BL Lacs and one quasar) in the EUVE all-sky survey. By searching the EUVE archive for sources near known extragalactic X-ray sources, Fruscione (1996) finds that 20 X-ray galaxies (12 Seyferts, one LINER, six Blazars, one quasar) are strong EUVE sources. High angular resolution EUV imaging of nearby galaxies does not exist.

The ultraviolet is very rich in spectral lines. These atomic and molecular lines are useful for deriving important astrophysical information. Hot (10,000–40,000 K) stars emit a large fraction of their radiation in the ultraviolet. Imaging studies in the far-UV and mid-UV detect hot stars associated with star-forming H II regions and young star clusters in spiral galaxies. Spectral studies of O VI (doublet at 1032 Å, 1038 Å) absorption in hot gas clouds by the FUSE satellite (launched on June 24th, 1999 and operational until October 18th, 2007) has given important diagnostic information about the intergalactic medium.

Far-UV and mid-UV images of ellipticals (e.g. M 32, the satellite galaxy of NGC 224/M 31; Figure 2.7) and spiral galaxy bulges has shown that the unexpected UV excess (first observed by the OAO series of satellites in the

Figure 2.7

UV image of M 32 by STIS on HST. The nucleus is at lower-right. The UV excess seen in ellipticals and bulges originates in the observed population of old, but hot, helium-burning stars.

Credit: NASA and T. M. Brown, C. W. Bowers, R. A. Kimble, A. V. Sweigart (NASA Goddard Space Flight Center) and H. C. Ferguson (STScI).

1970s) is not caused by recent massive star formation, but probably by low-mass, post-giant-branch stars.

The emission of quasars peaks around the UV region – "the big blue bump" is the well-known feature with a peak energy around the Lyman limit of 1216 Å (Risaliti and Elvis 2004). This peak is best described by thermal emission from accretion disk gas displaying a wide range of temperatures.

UV radiation is greatly attenuated by dust grains. These grains are very good absorbers of photons which have wavelengths equal to or smaller than the size of the grain. Many grains have characteristic sizes of 100 nm or more which means that interstellar dust absorbs UV radiation very efficiently.

2.4.4 Optical

3300 to 8000 Å

In normal galaxies optical emission is dominated by radiation from the photospheres of stars. Stars radiate in a similar fashion to a blackbody, with surface temperatures ranging from 3000 K (M dwarfs) to 40,000 K (O type). Our Sun is a G dwarf type spectral class with a

---

9 A rapid or dramatic brightening due to an explosive event, e.g. novae, or a flare.

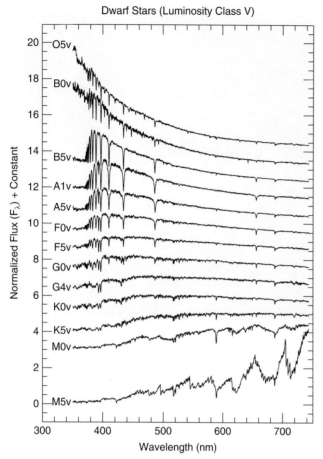

Figure 2.8

Optical spectra of dwarf stars from Jacoby, Hunter and Christian (1984) showing O to M spectral types.

Credit: Supplied by R. Pogge. Figure courtesy of G. Jacoby/NOAO/AURA/NSF. Reproduced by permission of the AAS.

surface temperature of 5800 K. Peak emission from stars with temperatures of 8700 K and 3625 K occurs at the limits of the optical region, 3300 Å and 8000 Å, respectively. Wien's law for a blackbody allows us to calculate the wavelength of maximum emission

$$\lambda_{\max} = \frac{0.0029 \text{ K m}}{T}$$

where $\lambda_{\max}$ is the wavelength of maximum emission in meters and T is the temperature of the object in K. The total amount of energy radiated by a blackbody is given by the Stefan–Boltzmann law

$$F = \sigma \, T^4$$

where F is the energy flux in joules $\text{m}^{-2} \text{ s}^{-1}$, σ is a constant (5.67×10^{-8} W m^{-2} K^{-4}), and T is the temperature in K. Based on these laws there are two key things to remember. Firstly, that temperature is inversely related

to $\lambda_{\max}$, hence if you double the surface temperature of a star, $\lambda_{\max}$ will halve. Secondly, the same temperature increase will increase the energy flux by a factor of 2^4 or 16.

Filters can be used to isolate a narrow range of optical emission. For example, observing a spiral arm, blue light (e.g. using a B filter with $\lambda_c \sim 4500$ Å) images will preferentially record radiation from young, hot stars whilst red light (R or I with $\lambda_c > 6000$ Å) images will be dominated by radiation from cooler, more evolved stars. Figure 2.8 shows dwarf star spectra (Jacoby, Hunter and Christian 1984) from O to M spectral types. Notice how the energy maxima increase to longer wavelengths in the progression from O (hot stars) to M (cooler) types.

Radiation from warm, 10^4 K gas, typically found in and near star-forming regions (also called H II regions), can also be detected in the optical region. This element of the ISM is usually detected by observing the recombination[10] emission line of singly ionized hydrogen (usually denoted by Hα) at 6563 Å. An example of this emission in the LMC is shown in Figure 2.9.

Other important diagnostic lines include the hydrogen Balmer series lines Hβ (4861 Å) and Hγ (4340 Å), and He I (5876 Å) and He II (4686 Å). Forbidden lines of ions such as the oxygen doublet [O III][11] (4959, 5007 Å) (Figure 2.10), the nitrogen doublet [N II] (6548, 6583 Å; surrounding the Hα line), the oxygen doublet [O II] (3726, 3729 Å) and the sulfur doublet [S II] (6716, 6731 Å) are also seen. H II regions are powered by UV radiation from nearby, hot stars. This radiation is absorbed by the gas and then re-emitted, mainly in the optical and IR regions.

Optical emission can be greatly attenuated by interstellar dust. Light is absorbed and scattered by dust particles or grains, and this attenuation is greater for shorter wavelengths. Hence, many B images in the atlas will show dramatic evidence of dust absorption in and near spiral arms, whereas in red images (e.g. R or I) the effect is less pronounced. Absorption and scattering can diminish the light from stars. Taken together astronomers refer to this as extinction. The extinction, A, is the difference between the observed magnitude and the magnitude in the absence of dust. Likewise, the color excess or reddening, E, is the difference between the observed color and the intrinsic color. The most

---

10 Recombination occurs when an electron is captured by an ion and energy in the form of photons is emitted corresponding to atomic energy levels.

11 [ ] indicates forbidden lines, where the "III" notation represents the doubly ionized species; similarly "II" is singly ionized, etc.

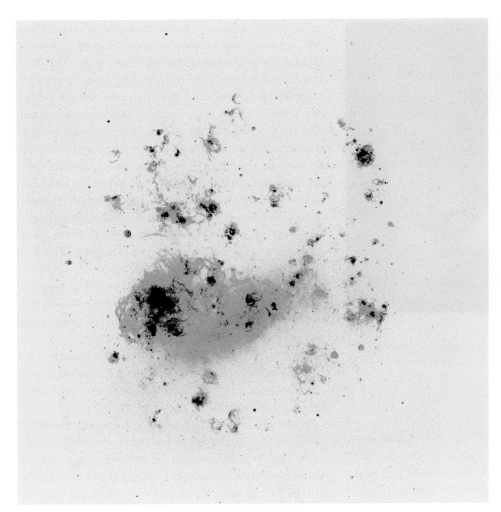

Figure 2.9

Hα image of the Large Magellanic Cloud.

Credit: C. Smith, S. Points, the MCELS Team and NOAO/AURA/NSF.

cited extinction is A_V in the optical V, and the color excess is

$$E(B - V) = (B - V)_{\text{intrinsic}} - (B - V)_{\text{observed}}$$

It is assumed that A tends to 0 at very long wavelengths, and

$$A_X = A_0 f(\lambda_X)$$

where A_0 is a constant and f is a theoretical function. Extinction curves for particular lines of sight can be determined. A_λ has a maximum in the far-UV whilst shorter wavelength X-rays can pass through dust grains, and much longer wavelength radiation refracts around the grains. A significant extinction "bump" or maximum exists at 217.5 nm which could be caused by graphite or PAHs. Several other features between 3.3 and 12 μm could be related to PAHs as well, as they have wavelengths of vibration modes in C–C and C–H bonds that are common in PAHs. In the far-IR, A_λ decreases with increasing wavelength as λ^{-1} but there is variation in A_λ particularly in the UV for different lines of sight.

The slope of the extinction curve near V in the optical is

$$\frac{A_V}{A_J R_V}$$

where J is 1.2 μm in the near-IR, and

$$R_V = \frac{A_V}{E(B - V)}.$$

Traditionally R_V has been taken to be 3.1 but it can range from ~3 (steeply increasing extinction into the UV) to ~5 (slowly increasing extinction into the UV). $E(B - V)$ is, not surprisingly, found to be proportional to the column density of interstellar hydrogen, N_H, since dust and cold gas seem to coexist in many environments:

$$E(B - V) = \frac{N_H}{5.8 \times 10^{25} \text{ m}^{-2}}$$

In general

$$E(B - V) \sim 0.53 \, (d/\text{kpc}) \quad \text{and} \quad A_V \sim 1.6 \, (d/\text{kpc})$$

for a line of sight of length d, in kpc.

Figure 2.10

Optical [O III] emission depicted as green in the Seyfert 2 galaxy NGC 1068/M 77. The nucleus is at the bottom-right and the cone is an artist's impression to show the opening angle of emission from the nuclear region. The image is ~1.5″ across.

Credit: Faint Object Spectrograph Investigation Definition Team, NASA.

Optical emission from AGN can be dominated by emission from their nuclei (Figure 2.10). The discovery of Seyfert galaxies was notable due to their extremely bright optical nuclei that made them resemble bright stars. Many AGN have very strong optical emission lines originating in gas clouds. Diagnostic information derived from AGN emission lines provide information about the origin and excitation of such lines via starburst or non-thermal processes. For example, ratios of line strengths (i.e. [O I]/Hα vs. [O III]/Hβ) help discriminate the origins of emission lines between H II regions powered by UV radiation from hot, young stars and various shock-front, high-energy excitation processes such as supernovae, jets or gas cloud–gas cloud collisions.

In many AGN both the emission lines and integrated stellar emission are swamped by much stronger synchrotron emission that increases its dominance on the total energy output at longer wavelengths.

2.4.5 Infrared

Near-IR: 0.8–7 μm; mid-IR: 7–25 μm; far-IR: 25–300 μm

Near-IR radiation can originate from stellar (i.e. cooler K- and M-type star) sources as well as being reprocessed by dust. Mid-IR and far-IR emission can be dominated

by radiation from interstellar dust grains such as carbon, hydrocarbons, silicates and polycyclic aromatic hydrocarbons (PAHs) heated by nearby stars. The IRAS wavebands in the mid-IR (12 μm and 25 μm) detect radiation which is dominated by non-thermal emission from small grains. The exact origin of this emission is still uncertain; however it may be a mixture of warm (~50 K) dust associated with star-forming regions, and cool (~20 K) dust associated with regions rich in atomic hydrogen, H I. The uncertainties are compounded by the unknown nature of the dust grain size and composition. Infrared "cirrus" is faint, wispy cloud-like emission (first discovered in IRAS images) seen above and below the plane of the Galaxy. This is believed to be emission from dust clouds associated with nearby H I clouds.

Strong IR emission is detected in starburst galaxies, interacting or merging galaxies and AGN. IR observations are particularly important in terms of determining the source of emitted radiation. The IR emission can be used to probe dusty areas such as the inner regions of AGN, as well as regions of high star formation (e.g. in starbursts and mergers) that are not visible or are heavily obscured in the optical.

Hydrogen recombination lines, especially Brα (4.05 μm) and Brγ (2.17 μm) of the Brackett series, and Pα (1.88 μm) of the Paschen series, are frequently observed. Other near-IR features include [Fe II] (1.64 μm), H$_2$ ($J = 1 \rightarrow 0$)[12] (2.12 μm), H$_2$ ($J = 2 \rightarrow 1$) (2.25 μm) and CO (2.34 μm).

2.4.6 Submillimeter

300 μm to 1 mm

Dust dominates the source of emission in the submillimeter region. Observations in this wavelength region have recently opened up due to innovations in instruments and detectors. Figure 2.11 shows an 850 μm image (Tilanus, van der Werf and Israel 2000) of the Whirlpool Galaxy, Messier 51. Spiral arm structure is clearly seen indicating the position of dusty star-forming regions.

Submillimeter observations of distant starburst galaxies and AGN are very important. As discussed more

---

12 Molecules move in space, at various speeds and directions. The energy and orientation of a molecule's tumbling motion is described as a rotational state, and these states are quantized. A molecule can spontaneously drop from its current energy state to the next lower one (i.e. a transition), converting the energy into a photon. The symbol ($J = 1 \rightarrow 0$) and others like it denote the particular energy level transition.

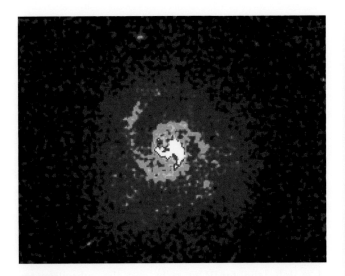

Figure 2.11

SCUBA image at 850 μm of the Whirlpool Galaxy, NGC 5194 or Messier 51. NGC 5195 is seen to the north.

Credit: JCMT/SCUBA. Image courtesy of R. Tilanus (JAC).

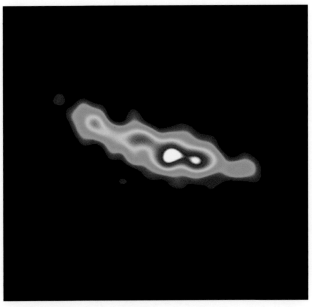

Figure 2.12

OVRO image at 92.0 GHz of NGC 3034/M 82.

Credit: OVRO. Image courtesy of E. Seaquist.

fully in Section 2.6, emission detected at a particular wavelength from distant galaxies originates at shorter wavelengths at the source due to cosmological expansion. Hence submillimeter observations of distant galaxies can detect source emission in the IR. For example, the detection at 850 μm of a galaxy with a redshift $z = 2.4$ will sample emitted radiation at 250 μm, or from the far-IR.

2.4.7 Radio

1 mm and longer wavelengths

Radio recombination[13] lines of hydrogen such as H41α (92 GHz, 0.33 cm), H29α (256.3 GHz, 0.11 cm), H27α (316.4 GHz, 0.09 cm) and H26α (354.5 GHz, 0.08 cm) are important diagnostics of ionized gas conditions such as temperatures and electron densities. Figure 2.12 shows continuum emission at 92.0 GHz near the hydrogen recombination line of H41α in the starburst galaxy, NGC 3034/M 82.

The measurement of CO is believed to directly indicate the mass of giant molecular clouds (GMCs) and can be used to estimate the amount of molecular hydrogen, H_2 (that is mostly cold, 10–20 K and hence not directly

observable – the 2.12 μm line is from warm H_2). Whilst the 2.6 mm CO ($J = 1 \rightarrow 0$) transition is most commonly observed, other lines such as 1.3 mm CO ($J = 2 \rightarrow 1$) and 0.88 mm CO ($J = 3 \rightarrow 2$) are studied as well. The Galactic factor, α, for the conversion between CO flux and H_2 column density[14]

$$\alpha = \frac{N_{H_2}}{S_{CO}} \text{ cm}^{-2} \text{ K}^{-1} \text{ km}^{-1}\text{s}$$

is given by (Omont 2007) as

$$\alpha = (1.8 \pm 0.3) \times 10^{20} \text{ cm}^{-2} \text{ (K km s}^{-1})^{-1}$$

for large molecular clouds away ($|b| > 5°$) from the Galactic plane and the molecular mass M_{mol} and CO flux S_{CO} are related by

$$M_{mol} = 1.61 \times 10^4 \, D_{Mpc}^2 \, S_{CO} \text{ M}_\odot.$$

Theoretical studies suggest that α could be a strong function of metallicity, density and excitation temperature.

Radio continuum emission can consist of non-thermal synchrotron radiation. Synchrotron radiation originates from old ($> 10^7$ yr) relativistic electrons which have typically travelled significant distances from their parent

---

13 Radio recombination lines occur via transitions of electrons between two energy states with very high quantum number n. These lines are named after the atom, the destination quantum number and the difference in n of the transition (α for $\delta n = 1$, β for $\delta n = 2$, etc.). An example is H41α (transition from $n = 42$ to $n = 41$ in hydrogen).

14 Column densities indicate the areal density of a given species, usually quoted in atoms cm^{-2}. N_n is the line integrated density of atoms in the nth state. Therefore N_{H_2} is the column density of molecular hydrogen.

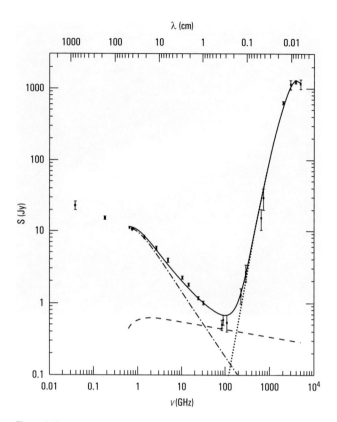

Figure 2.13

Radio/submillimeter/far-IR spectrum of NGC 3034/M 82. The observations (data points) are fitted by a model (solid line) that consists of synchrotron (dot-dash line), free–free (dashed line) and dust (dotted line) components.

Credit: with permission, from the Annual Review of Astronomy and Astrophysics, Volume 30 © 1992 by Annual Reviews www.annualreviews.org

SNRs. Powerful synchrotron emission can also be observed as core, jet or lobe emission due to nuclear activity in AGN (as described in Section 1.5.5). Some of the continuum radiation can also be thermal radiation from star-formation or warm gas regions due to free–free[15] interactions of electrons. Carilli *et al.* (1991) investigated the energetics of the radio emission in Cygnus A and confirm the jet model for powerful radio galaxies. A synchrotron aging process occurs in which energetic particles are made at the radio hotspots, expand into the radio lobes and lose energy via the synchrotron process.

The complexity of emission in the starburst galaxy NGC 3034/M 82 is shown in the radio/submillimeter/far-IR spectrum depicted in Figure 2.13. The radio region

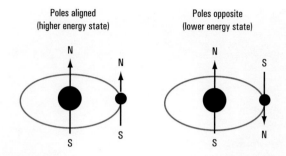

A 21 cm photon is emitted when poles go from being aligned to opposite (a spin flip).

Figure 2.14

Emission of 21 cm radiation from the hydrogen atom.

Credit: T. Herter.

is at left, the far-IR at right and the submillimeter region between wavelengths of 0.1 and 0.03 cm. The observed spectrum (data points) is well represented by the solid line model that is composed of three different model emission mechanisms. This combined model is made up of synchrotron (dot-dash line), free–free (dashed line) and dust (dotted line) components. The synchrotron radiation dominates at wavelengths greater than 10 cm. The starburst-induced free–free emission dominates between 30 and 200 GHz. Re-radiated emission from dust dominates the spectrum at frequencies greater than 200 GHz or wavelengths less than 1 mm in the submillimeter and far-IR regions. The complete spectrum of NGC 3034/M 82 is shown in Figure H.4 in Appendix H.

Radiation from atomic hydrogen (H I) is emitted at the radio wavelength of 21 cm.[16] This radiation occurs when the hydrogen atom changes from a high to low (preferred) energy state, as its electron changes its spin direction. Figure 2.14 depicts the relative spin directions of the proton and electron. This emission is called line radiation, because of its narrow wavelength distribution. The detection of H I shows the neutral (non-ionized) cold gas distribution (Figure 2.15) and identifies gas motions based on the detected wavelength of the 21 cm line. The H I mass of a galaxy can be calculated by

$$M_{\mathrm{HI}} = 2.36 \times 10^5 \, D_{\mathrm{Mpc}}^2 \, \Sigma \Delta V \, \mathrm{M}_\odot$$

where D_{Mpc} is the distance to the galaxy in Mpc, and $\Sigma \Delta V$ is the integrated line flux in Jy km s^{-1}, where Jy is Jansky, the unit of flux.

---

15 When an electron collides with an atom or ion, and quantum mechanically emits or absorbs a photon.

16 Astronomers refer to this emission line as 21 cm – more accurately its vacuum wavelength is 21.11 cm and frequency is 1420.41 MHz.

Figure 2.15

H I radio image of the Small Magellanic Cloud. This emission shows the distribution of cold, atomic hydrogen gas. RA spans 0 h 25 m (right, bottom) to 1 h 40 m (left, bottom). Dec. spans −70° 20′ (top) to −75° 10′ (bottom). The H I map was observed with ATCA, over 8 × 12 hour observing periods and has a spatial resolution of 98″.

Credit: S. Stanimirovic, L. Staveley-Smith and CSIRO.

The H I mass of disk (spiral) galaxies normalized by their optical luminosity, $M_{\rm HI}/L_{\rm B}$, tends to increase in a systematic way from $\sim$0.05 $M_\odot/L_\odot$ for Sa spirals, to $\sim$1 $M_\odot/L_\odot$ for Magellanic irregulars (Sm and Im classes). Strong H I in ellipticals is not common, though H I detections suggest that such cold gas may originate from gas-rich galaxies that have merged with the elliptical.

2.5 Caveat #1: Mass versus light and dark matter

I think many people initially wished that you didn't need dark matter. It was not a concept that people embraced enthusiastically. But I think that the observations were undeniable enough so that most people just unenthusiastically adopted it. VERA RUBIN

This atlas shows images which depict various types of radiation originating in or near galaxies. However, when considering the total amount of matter in these objects, observations only directly detect radiation from a small fraction of the total mass of each galaxy. Measurements of the velocities of gas (e.g. neutral hydrogen) as a function of radius in spirals, and the detection of hot, 10^7 K, gas around ellipticals, imply that dynamically inferred galaxy masses are a factor of $\sim$5 or greater than the masses deduced from the combined luminosities of the stars and gas. In fact the atlas galaxy NGC 2915 (page 105) has a dark matter content probably a factor of 50 greater than its luminous matter, suggesting that dark matter makes up $\sim$98% of the total galaxy mass.

In a sense observations only detect the tip of each galactic "iceberg", with the majority of the mass of each galaxy remaining unseen. The first quantitative study of this unseen matter was carried out early in the twentieth century (Oort 1932) by observing the vertical motions of stars in our Galaxy. The local mass density near the Sun was initially determined to be $\sim$0.15 $M_\odot$ pc^{-3}, with dark matter representing 40% of the total local mass, although by the 1980s this was shown to be an overestimate.

Early observations of the Andromeda Galaxy also showed discrepancies between dynamical and luminosity based masses. The rotation curve of NGC 224/M 31 (Babcock 1939) implied a *global* mass-to-light ratio (M/L) of $\sim$14 (corrected for the present-day distance to NGC 224/M 31), which was a factor of 10 higher than that implied at the nucleus. The discrepancies were even larger in the case of galaxy clusters. Observations of clusters showed individual galaxies with much larger radial velocities than could be accounted for assuming a gravitationally bound cluster. Smith (Virgo Cluster; Smith 1936) and Zwicky (Coma Cluster; Zwicky 1937) detected a large mass discrepancy in these nearby clusters. In a landmark study of disk galaxies Ken Freeman (1970) commented on the rotation curves of M 33 and NGC 300 and noted that they did not show an expected Keplerian velocity decline[17] with their optical radii:

> *For NGC 300 and M 33 ... there must be in these galaxies additional matter which is undetected ... Its mass must be at least as large as the mass of the detected galaxy, and its distribution must be quite different from the exponential distribution which holds for the optical galaxy.* KEN FREEMAN

Freeman had re-kindled the question of dark matter, this time though *in galaxies*, and was also the first to speculate on its structure. In the mid-1970s Rubin and colleagues observed rotation curves of spirals that were flat or even rising to large radii that finally brought the problem to wide attention.

Theoretical models suggest large, massive dark matter halos exist around most if not all galaxies. The size of the halos may be very large. For example, whilst the distance between the Galaxy and NGC 224/M 31 is $\sim$700 kpc and if the luminous diameters of both galaxies are $\sim$40–50 kpc (but see below), a large fraction of the intervening distance could be taken up with their individual dark matter halos. The dark halos may even overlap.

Deep exposure imaging (McConnachie *et al.* 2009), Figure 2.16, around NGC 224/M 31 and towards its Local Group neighbor, NGC 598/M 33, has shown stellar structure extending 150 kpc from NGC 224/M 31 and overlapping with stellar light extending 50 kpc from NGC 598/M 33. It is likely that stars exist between both galaxies and could exist as far away from parent galaxies as the virial radius of their dark matter halos ($\sim$300 kpc for NGC 224/M 31).

The virial theorem relates the total kinetic energy of a self-gravitating body due to the motions of its constituent parts, K, to the gravitational potential energy, U, of the body such that $2K + \mathrm{U} = 0$. For gravitationally bound galaxies in equilibrium, and with some assumptions, the relationship

$$M = \frac{2v^2 R}{G}$$

follows, where M is the total mass of the galaxy, v is the mean velocity (the sum of the rotation and velocity dispersion) of stars in the galaxy, G is Newton's gravitational constant and R is the effective radius (size) of the galaxy. Defining the virial radius is somewhat more complicated. For an unlikely occurring system with equal masses, the virial radius r_v is the inverse of the average inverse distance between particles, r. For a more likely system of i particles with different masses, m, and total mass M, the definition is

$$\frac{M^2}{r_\mathrm{v}} = \sum_i \sum_{j,j \neq i} \frac{m_i m_j}{\left| r_i - r_j \right|}.$$

For simplicity r_v can be thought of as the radius of a sphere, centered on a galaxy or a galaxy cluster, within which virial equilibrium holds. In practical terms it is often approximated as the radius within which the average density is greater, by a specified factor, than the critical[18] density

$$\rho_\mathrm{crit} = \frac{3H_0^2}{8\pi G}$$

where H_0 is the Hubble constant.[19] A common choice for the (over) density factor to describe r_v is 200 (early simulations suggested that the radius at a density factor of 178 is close to r_v) in which case r_v is approximated as r_{200}.

---

17 If the total mass converges at some radii, the rotation curve would behave as $V \propto 1/r$, in which velocities at large r would eventually decline.

18 The mass density below which the Universe is open (positive space curvature), and above which the Universe is closed (negative space density) is the critical density. It is approximately $\rho_\mathrm{crit} = 1.0 \times 10^{-26}$ kg m^{-3}.

19 The linear relationship between a galaxy redshift and its distance is commonly specified as $H_0 = 100h$ km s^{-1} Mpc^{-1}, the Hubble constant. The dimensionless quantity h, until recently, was taken to be between 0.5 and 1.0, but now appears to be close to 0.72.

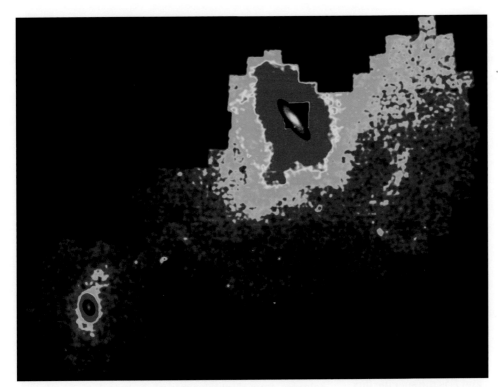

Figure 2.16

Stellar density map of the Andromeda–Triangulum region. A projection of the stellar density distribution with scale images of the disks of NGC 224/M 31 (top-right) and NGC 598/M 33 (bottom-left) overlaid.

Credit: A. McConnachie. The small images of M 31 and M 33 courtesy T. A. Rector, B. A. Wolpa and M. Hanna (NRAO/AUI/NSF and NOAO/AURA/NSF). Reprinted by permission from Macmillan Publishers Ltd: Nature, vol. 461, p. 66, copyright (2009).

The extended stellar distributions comprise both individual stars lost from each system during previous nearby interactions and the destroyed remnants of dwarf galaxies due to the tidal field of each galaxy. These observational signatures are consistent with galaxy growth inside dark matter halos that grow via accretion and mergers events.

Candidates for the origin of dark matter, both of baryonic[20] (e.g. black holes, brown dwarfs and white dwarfs) and non-baryonic (e.g. neutrinos with mass) forms, are numerous. Current models of dark matter and computer simulations of large-scale structure compared to observed structure suggest that the majority of dark matter is "cold". Cold dark matter (CDM) would comprise slow-moving, massive particles. In comparison, hot dark matter (HDM) would consist of fast-moving, light particles. Experiments suggest that neutrinos could have mass and could therefore be a candidate for HDM, but are probably not massive enough to be a dominant part of dark matter. Primack (2009) presents an overview of dark matter in relation to galaxy formation in the context of the ΛCDM "double dark" standard cosmological model. "Darkness" appears to rule the

Universe. CDM plus "dark energy" (denoted by Λ; the unknown force or property of the vacuum driving the acceleration of the Universe) make up 95% of the cosmic density. The search for the origin of dark matter continues.

2.6 Caveat #2: Looking back to the beginning

Detecting galaxies at cosmologically large distances introduces several observational biases.

Firstly, because of the expansion of the Universe, their emitted radiation is detected at longer, redshifted[21] wavelengths. For example, the optical Hα emission line (emitted at $\lambda = 6563$ Å, in the rest-frame) in a galaxy at redshift $z = 1$ will be detected by a telescope at $\lambda = (1 + z) \times 6563$ Å $= 13{,}126$ Å (1.3126 μm in the near-IR).

Sources observed at different redshifts are sampled at different rest-frame wavelengths. Photometry is performed with a fixed bandpass filter, so the effective

---

20 A massive elementary particle made up of three quarks. Neutrons and protons are baryons.

21 The vast majority of galaxies are redshifted. A few, nearby galaxies such as NGC 224/M 31, that are influenced by local gravitational fields, are approaching us, and their radiation is blueshifted to smaller wavelengths.

width of the bandpass will change with different source redshifts. The correction for this effect, which transforms a measurement of a source at a redshift z, into a standard measurement at redshift zero or the rest-frame, is called the "K correction". It is dependent on galaxy type and redshift.

For a source observed with an apparent magnitude m_Y, through the photometric bandpass Y, which has an absolute magnitude M_C in an emitted frame bandpass C, the K correction $K_{CY}(z)$ is defined by

$$m_Y = M_C + DM + K_{CY}(z)$$

where DM is the distance modulus of the source, defined as

$$DM = 5 \log_{10} \frac{D}{10 \text{ pc}}$$

and D is the source distance in pc. For galaxies that can be described by a power law

$$F_\nu \propto \nu^{-\alpha}$$

where α is the power-law index and F is the specific flux, then

$$K_{CY}(z) = 2.5 \, (\alpha - 1) \, \log_{10}(1 + z).$$

Secondly, at very large distances the inverse-square law of radiation propagation fails as the radiation surface area is no longer described by a wavefront on a normal sphere surface. A consequence of this is that the surface brightness of an extended, distant object is redshift dependent and scales as $(1 + z)^{-4}$. This is known as cosmological or Tolman dimming. The large exponent ensures this effect is quite drastic even at $z \sim 2$ when the cosmological dimming factor of surface brightness is 81.

Thirdly, there is an age or evolutionary effect. Due to the substantial light travel times to distant galaxies their radiation is detected when they were at much younger ages. Stellar populations evolve, thus as more and more distant galaxies are detected, the radiation from successively younger and younger populations is recorded.

A long-exposure image by HST WFPC2 in the constellation of Ursa Major is known as the Hubble Deep Field – North (HDF-N). Four filters (F300W, F450W, F606W and F814W), spanning the mid-UV through optical to near-IR region were combined to give a "true-color" view of the distant Universe. A sight line out of our Galaxy with a low density of foreground Galactic stars supplied a clear view. The majority of objects in the image are distant

galaxies, some with $z \sim 3$, implying light travel times of ~ 10 Gyr. Ignoring stellar population evolution, for galaxies at $z = 3$, observed in the F814W filter ($\lambda = 8140$ Å), observations detect emitted (at the galaxy) radiation of $\lambda = (8140 \text{ Å}/1 + z) = 2035$ Å, which is in the mid-UV. Such effects have to be taken into account when observing distant objects.

Since detected radiation is redshifted by significant amounts from distant galaxies it is worth observing these galaxies at IR and longer wavelengths. Many galaxies are strong emitters in the optical and near-IR regions which will help in the detection of distant sources. Such wavelength regions will also contain re-emitted radiation from dusty sources. Since many distant, young galaxies and starburst galaxies will have high dust contents due to intense star formation, the submillimeter is an excellent region to utilize. Figure 2.17 is a JCMT SCUBA image of the HDF-N at 850 μm.

After 50 hours of integration time this image represents one of the deepest submillimeter images ever taken. Five discrete sources have been identified with galaxies, four of which are likely to be galaxies with redshifts in the range $2 < z < 4$. The submillimeter results

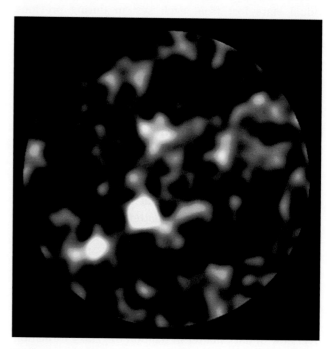

Figure 2.17

JCMT SCUBA 850 μm image of the Hubble Deep Field – North. The radius of the field is 100″. The field is centered at 12 h 36 m 51.2 s, +62° 12′ 52.5″ J2000. Five sources have been identified with galaxies. Data from Hughes *et al.* (1998).

Credit: Reprinted by permission from Macmillan Publishers Ltd: Nature, vol. 394, p. 241, copyright (1998).

Figure 2.18

Hubble Ultra Deep Field – infrared. A 48 hour integration by HST WFC3 in the constellation of Fornax. Three filters (F105W – blue, F125W – green and F160W – red) are combined. The majority of objects are distant galaxies. The image is ~2.4' across.

Credit: NASA, ESA, G. Illingworth (UCO/Lick Observatory and UCSC), R. Bouwens (UCO/Lick Observatory and Leiden University) and the HUDF09 Team.

indicate that the star-formation rates in these distant galaxies are about five times higher than that indicated by the UV properties of the HDF-N galaxies. This result highlights the importance of the submillimeter region as an accurate indicator of star formation for distant galaxies.

Not to be outdone, CXO has also observed this region of sky. About 556 hours (23 days!) of observations were combined in the Chandra Deep Field North image. Many of the objects detected are AGN with SMBHs. These data are being used to determine when such black holes form and how they evolve. The data show that SMBHs are quite rare at the very earliest times in the Universe suggesting that they need time to grow by feeding on gas and stars. Detecting X-ray photons is difficult! The faintest sources in the image produced only *one* X-ray photon every four days.

The power of observing distant galaxies at longer wavelengths is dramatically shown in Figure 2.18 which is a long exposure of the Hubble Ultra Deep Field using HST WFC3 near-IR filters with wavelength centers of 1.05, 1.25 and 1.55 μm. Comparison should be made with Figure 1.18 which was a predominantly optical image of the same area of sky. The WFC3 near-IR image shows numerous small, red objects, not seen in the ACS optical-based image; many of these are distant galaxies, seen ~600 million to 1 Gyr after the Big Bang.

Figure 2.19 is an image of galaxies and clusters of galaxies in the making. It shows the sky in the microwave region, expressed as minute fluctuations in temperature above and below (red and blue, respectively) the cosmic microwave background (CMB) temperature of 2.73 K.

The CMB all-sky emission at 2.73 K was detected and understood in 1965 as remnant radiation from the Big Bang start to the Universe. The Big Bang "fireball" has expanded and cooled after 14 Gyr and this remnant radiation has the temperature and blackbody characteristics consistent with such a hot beginning. What exactly is seen in Figure 2.19? Our early Universe consisted of a dense, hot "soup" of subatomic particles and extremely high-energy photons interacting with one another. As the fireball expanded the density and temperature decreased. Small, weak density variations evolved, and altered the temperature of the photons. Lower density regions are the temperature hot-spots in the CMB and higher density regions correspond to colder regions.

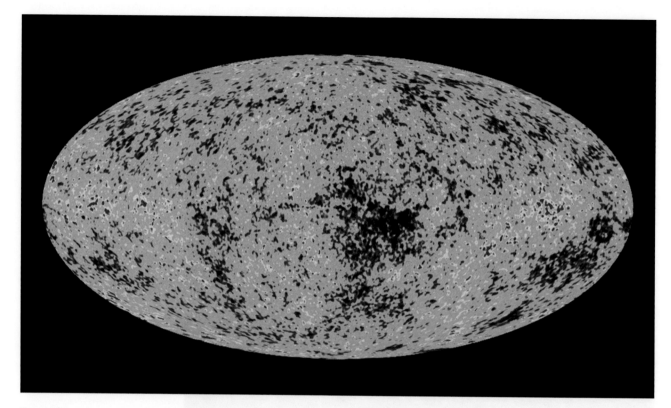

Figure 2.19

The all-sky cosmic microwave background observed by WMAP.

Credit: NASA/WMAP Science Team.

About 300,000 years after the Big Bang, the temperature had reduced to 3000 K which was cold enough so that subatomic particles (i.e. protons, neutrons, electrons) could combine to form atoms. Photons could then travel without significant scattering or absorption and the Universe became "transparent". The last interactions of photons with matter, in particular electrons, occurred at this time (300,000 years corresponds to a redshift of $z \sim 1000$) and this is what is observed as the CMB. It is called the last scattering surface and it is the signature of the Universe at the time the first structures of matter formed. Gravity then took over and galaxies eventually formed. Hence the CMB is an important link between the hot, smooth early Universe devoid of galaxies and the much cooler, lumpy Universe full of galaxies today.

In the early 1990s the Cosmic Background Explorer (COBE) satellite discovered small (10 parts in a million) differences in the CMB temperature across the whole sky. These deviations are very important since they are needed if structures such as clusters of galaxies and individual galaxies are to form. However, the angular resolution of $\sim 7°$ for COBE was not sufficient to determine the smallest sizes of these deviations. In late 1998 the Balloon Observations of Millimetric Extragalactic Radiation and Geophysics experiment, or BOOMERANG, was launched in the Antarctic and circumnavigated the continent at an altitude of ~ 37 km for about 10 days. BOOMERANG delivered observations with an angular resolution of $\sim 0.2°$ and sampled over 3% of the sky.

The Wilkinson Microwave Anisotropy Probe (WMAP) was launched on June 30th, 2001 and has produced all-sky microwave maps (Figure 2.19) with BOOMERANG-like angular resolution ($\sim 0.3°$) providing the best yet information about early Universe structure formation and evolution. It has provided an estimate of the age of the Universe of 13.69 ± 0.13 Gyr as well as the baryon, dark matter and dark energy density, the Hubble constant, H_0 and the total neutrino mass (Dunkley *et al.* 2009).

2.7 Caveat #3: Observational bias

All scientific measurements have an inherent bias. The selection of a particular telescope, instrument and detector combination will bias astronomical detections, whether it be due to the diameter of the telescope mirror, the wavelength of detection, the length of the observation, or the efficiency and characteristics of the detector.

A clear-cut example of sample bias is found in most surveys of distant galaxies. Galaxies close to the detection limit (i.e. at the faint limit of the survey) will be more luminous on average than those at a brighter limit. Simply put, towards fainter levels of detection, which usually means detecting more distant galaxies, it becomes progressively more difficult to detect intrinsically faint objects, and the sample gets skewed towards more luminous objects. This is known as "Malmquist bias" and it affects all galaxy surveys that are flux limited.

Galaxies are diffuse, extended objects. They are detected against the competing brightness of the night sky, which is typically 23 magnitudes per square arcsecond (mag arcsec^{-2}) in the optical B. Depending on your observational set-up, below a certain fraction of the night sky brightness no galaxy will be detected. A "censorship of surface brightness" exists and until recently most studies of galaxies would tend to concentrate on the brighter surface brightness examples. There is, however, increasing efforts to detect and study low surface brightness (LSB) galaxies. Although there is no formal convention for defining an LSB galaxy, they typically have central surface brightness fainter than 23 B mag arcsec^{-2}.

Malin 2 (page 108) is an example of a LSB galaxy. LSBs are made up of a variety of galaxy types, from giant, gas-rich disk galaxies (i.e. Malin 2) to dwarf spheroidal galaxies. The extreme ultra-gas-rich LSB galaxies have similar properties to small groups or clusters of galaxies, suggesting they have experienced a different evolution history than their high surface brightness cousins.

2.8 Galaxy research and multiwavelength observations

There are numerous areas in galaxy research that owe their existence to the power of panchromatic observations. Many of these areas are still being pursued as open areas of research and the following sections describe a few of these areas.

2.8.1 A unified scheme of active galaxies

Attempts have been made to show that the numerous AGN classifications and properties can be explained by a "unified scheme". This scheme (Antonucci 1993; Urry and Padovani 1995) suggests that the differences seen in

AGN can be accounted for by observing the same type of active galaxy along different lines of sight through different orientations of a similar central structure.

In the unified scheme (Figure 2.20) every AGN contains an obscuring dust and gaseous doughnut-like torus around a central supermassive black hole (SMBH) that has a surrounding thin gaseous accretion disk. Highly collimated relativistic jets are produced at right angles to the plane of the dusty torus. Observing the nucleus via a line of sight that intersects the plane of the torus would obscure the active core of the accretion disk and SMBH. It would appear as a radio galaxy, for an elliptical host galaxy, or as a Sy 2, for a spiral host galaxy.

Up to a distance of several kpc from the nucleus are the gaseous "narrow line regions" (NLR) excited either by photoionization from the UV continuum of the central source or by shock excitation related to the jets. The gas clouds visible would be predominantly far away from the nucleus, with low density, slowly moving, and having small linewidths ($\sim$500 km s^{-1}). Alternatively, observing

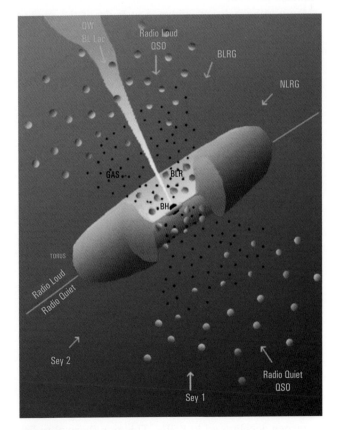

Figure 2.20

A schematic representation of the unified model of active galaxies. Various sight lines into the AGN are indicated. BLRG stands for broad line region galaxy; NLRG stands for narrow line region galaxy.

along the direction of the relativistic jet allows a clear view of the nucleus, so it would appear as a blazar, or a Sy 1 for a spiral host. Gas orbiting close to the SMBH is photoionized, producing the Doppler-broadened emission lines characteristic of the "broad line region" (BLR). The gas clouds are dense with typical speeds of 5000 km s^{-1}, thus having wide or broad linewidths. Intermediate viewing angles could show an AGN with properties of a broad lined Sy 1, quasar or radio galaxy.

Shortfalls in the unified scheme can be explained by variations in the properties of the structural components. For example, variations in black hole mass, the size and mass of the dust torus, galaxy gas content (i.e. possible fuel for a black hole), and the dynamical state of the host galaxy (Dopita 1997) could easily exist. Ho (2008) also promotes the idea that LINERs and other low-luminosity AGNs may not be simple scaled-down versions of their higher luminosity relatives. Their central engines may be qualitatively different. Whilst the unified scheme does not explain all observed AGN characteristics it is the best model at present.

Supporting indirect evidence in favor of the unified scheme comes from observations of the cores of nearby elliptical galaxies. Large-velocity gas motions close to the nuclei in many galaxies indicate masses of $\sim 10^9$ M$_\odot$ inside an area not much larger than our Solar System that could be explained by the presence of a SMBH.

2.8.2 The far-IR radio correlation

A correlation was found between the far-IR and radio continuum strengths of normal spiral galaxies (van der Kruit 1973). This initial finding was confirmed by the IRAS satellite. Using IRAS 60 and 100 μm flux densities, the far-IR flux of normal star-forming galaxies is strongly correlated with their 20 cm radio flux. The far-IR-radio correlation (Figure 2.21) ranges over five orders of galaxy luminosity. What is remarkable is that this very strong and universal correlation seems to originate from the same stellar population, high-mass stars, giving rise to two very different emission mechanisms, one thermal, the other non-thermal. It is now established that the far-IR emission is thermal and originates in dusty H II regions heated by high-mass stars. The 20 cm luminosity is mainly synchrotron radiation from relativistic electrons accelerated in SNRs. This non-thermal radiation dominates any radio free–free (thermal) emission that may occur in ionized gas regions. For example, Figure 2.13 shows the dominance of synchrotron over

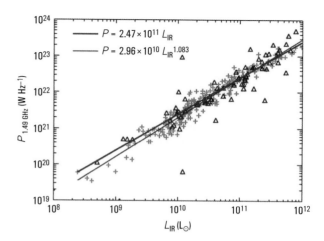

Figure 2.21

1.49 GHz radio power versus far-IR luminosity for IRAS galaxies.

Credit: Data kindly provided by E. Dwek; original form in Dwek and Barker (2002). Reproduced by permission of the AAS.

free–free emission at 20 cm in NGC 3034/M 82. The SNRs are, of course, direct evolutionary consequences of the same high-mass stars (Condon 1992).

Dopita (2005) discusses the correlation for starburst galaxies and summarizes that if the synchrotron electrons are short-lived compared to the starburst phase time-scale, the synchrotron emissivity relates directly to the supernova rate which in turn should be proportional to the star-formation rate.

The linearity of the far-IR radio correlation has been questioned. Bell (2003) has compared the far-UV derived star-formation rates with that from the far-IR and found that the far-IR traces most of the star formation in luminous $\sim$L* galaxies but traces only a small fraction of the star formation in faint $\sim$0.01L* galaxies. Since the far-IR-radio correlation is very close to linear at low luminosities, Bell (2003) suggests that the non-thermal radio flux is also decreased, which conspires to give such a linear correlation. This work is difficult for a number of reasons. Firstly, small numbers of low-luminosity galaxies (usually selected or enhanced across non-homogeneous samples) are used in many such studies. Secondly, there are potential problems in relating UV-derived to other star-formation rate indicators because of the different stellar time-scales probed by the UV relative to other tracers of ionizing photons. Calzetti *et al.* (2007) study star formation based on mid-IR emission of local galaxies and show that their viability as SFR indicators is subject to a number of caveats. The most robust star-formation indicator combines the observed Hα and 24 μm luminosities as probes of the total number of ionizing photons present in a region.

2.8.3 A non-universal IMF

The initial mass function (IMF), $\xi(M)$, describes the mass distribution of stars formed in a particular region. It takes the form of a power law

$$\xi(M) = cM^{-(1+x)}$$

with $\xi(M)$ existing over a range of stellar masses; these limits have been $M_{lower} = 0.1$ M$_\odot$ to $M_{upper} = 125.0$ M$_\odot$ in three of the most popular IMFs (Salpeter 1955 with a slope of $x = 1.35$; Miller and Scalo 1979; Scalo 1986). Until recently it was assumed, possibly for simplicity, and somewhat naively, that the IMF was universal.

The results of Meurer *et al.* (2009) now challenge this assumption. Using a large sample of H I selected galaxies, the ratio of Hα to far-UV flux is found to correlate with the surface brightness in Hα and the optical R. It is found that low surface brightness (LSB) galaxies have lower Hα to far-UV flux ratios than high surface brightness galaxies, and do not replicate the ratios derived from popular star-formation models using IMF parameters (as above). The authors suggest the correlations are systematic variations of the upper stellar mass limit and/or slope (x) of the IMF at the high-mass end. Simply stated, low-luminosity galaxies have less massive stars than higher luminosity galaxies. Yet it appears that the surface brightness drives IMF variations more than luminosity or mass. These results imply that the rate of star formation derived is highly sensitive to the indicator used in the measurement, as mentioned in the previous section.

2.8.4 Gamma ray bursts

A GRB detected on May 8th, 1997 (GRB970508) by the BeppoSAX satellite was the first GRB for which a secure distance could be estimated. An associated optical transient was detected 5.8 hours after the discovery (Djorgovski *et al.* 1997) and its brightness decline was consistent with relativistic "fireball" models of extreme energy events. Spectroscopy of the optical transient (Metzger *et al.* 1997) showed absorption lines due to a galaxy at a redshift of $z = 0.84$. The absence of Lyman-α (hydrogen) absorption in the spectrum implied that the optical transient had a redshift $0.84 < z < 2.3$. At the minimum redshift of $z = 0.84$, this GRB would have a total luminosity of 7×10^{51} erg s^{-1}. To put this in perspective, the luminosity of the *brief* GRB970508 event

was equivalent to the optical luminosity of more than 10^8 galaxies similar to the Galaxy. A more recent burst, GRB090423, detected by the Swift satellite, is associated with a galaxy at $z = 8.2$. This implies that the GRB is being detected only after $\sim$5% of the age of the Universe has expired.

Theoretical models have been proposed to explain the energy source and emission mechanism of these bursts. Candidate models include neutron star–neutron star or black hole–black hole mergers (or a combination of the two objects) and hypernovae that are extremely massive stars undergoing supernova collapse to a black hole. These models usually rely on some sort of preferential beaming of radiation along our line-of-sight to achieve the observed extreme luminosities.

Observations have been strengthening the case for a GRB–supernova link. CXO detected a spectrum of hot gas moving at $0.1c$ for GRB020813 containing numerous elements commonly seen in the ejecta of supernova explosions. The data appear to support the hypernova model in which shock waves in jets emanating from the black hole region produce gamma rays and X-rays. The CXO observations detect the radiation from the jets interacting with the expanding shells of ejected gas.

2.8.5 Magnetic fields in galaxies

Large-scale magnetic fields exist in spiral (grand design and flocculent[22]), irregular and dwarf irregular galaxies and are coherent over scales as large as 1 kpc. Their origin and evolution is directly linked to induction effects in the ionized ISM. These fields are likely driven by a large-scale dynamo action that can exist even whilst other hydromagnetic effects occur either globally (i.e. spiral arm related density wave propagation; bars) or locally (i.e. galactic fountains, shocks) (Beck *et al.* 1996; Krause 2003).

Continuum radio observations allow the detection of galactic magnetic fields via synchrotron emission. The intensity of linearly polarized emission measures the magnetic field (Figure 2.22), after correction for Faraday rotation and depolarization.[23] Observations at two or more frequencies allow the correction to be computed.

---

22 Flocculent spirals have patchy disk structures rather than clear spiral arms.

23 The rotation of the plane of polarization of a linearly polarized wave propagating through a magnetized dielectric medium is known as Faraday rotation (discovered by Michael Faraday in 1845). The rate at which the plane of polarization rotates is proportional to the product of the electron number density and the parallel magnetic field strength.

Figure 2.22 (opposite)

The magnetic field (vectors) and radio continuum (contours) of NGC 5194/M 51 overlaid on the optical.

Credit: radio continuum (white contours) and magnetic field (yellow vectors), MPIfR A. Fletcher, R. Beck; Optical: NASA/Hubble Heritage, STScI; Graphics: Sterne and Weltraum.

Regular magnetic fields align with spiral arms but are strongest in the inter-arm regions. Total field strengths are on average 8 μG[24] with some inter-arm regions reaching 20 μG. It is thought that higher turbulent gas velocities in the arm regions decreases the dynamo strength promoting greater field strength in the inter-arm areas. Magnetic fields are also observed in barred galaxies, and total field strengths correlate with the length of the bar. Edge-on galaxies display magnetic fields that are mainly parallel to their disks except in some galaxies with strong star formation and galactic winds.

2.9 Additional reading

The listing is not intended to be comprehensive but will complement the text. Advanced texts or papers are indicated with the † symbol.

J. J. Condon, 1992, Radio emission from normal galaxies, *Annual Review of Astronomy and Astrophysics*, **30**, 575.

L. C. Ho, 2008, Nuclear activity in nearby galaxies, *Annual Review of Astronomy and Astrophysics*, **46**, 475.

R. C. Kennicutt, Jr., 1998, Star formation in galaxies along the Hubble sequence, *Annual Review of Astronomy and Astrophysics*, **36**, 189.

I. Robson, 2009, The submillimetre revolution, *Experimental Astronomy*, **26(1-3)**, 65.

F. D. Seward, and P. A. Charles, 2010, *Exploring the X-ray Universe*, 2nd edn (Cambridge University Press, Cambridge).

B. T. Soifer, G. Helou and M. Werner, 2008, The Spitzer view of the extragalactic Universe, *Annual Review of Astronomy and Astrophysics*, **46**, 201.

†M. S. Longair, 2010, *High Energy Astrophysics*, (Cambridge University Press, Cambridge).

†D. E. Osterbrock, 1989, *Astrophysics of Gaseous Nebulae and Active Galactic Nuclei* (University Science Books, Mill Valley, CA).

†J. A. Peacock, 1999, *Cosmological Physics* (Cambridge University Press, Cambridge), Chapters 12–14.

†L. Spitzer, 1998, *Physical Processes in the Interstellar Medium* (Wiley-Interscience, New York).

---

24 G stands for gauss, the cgs unit of measurement of a magnetic field, named after Carl Friedrich Gauss (1777–1855). The Earth's magnetic field measures about 0.5 gauss, and the surface of a neutron star has a field of about 10^{12} gauss.

PART 3

A view from the inside: The Galaxy

Iɴ this part, we discuss the properties of the Galaxy center, and finish with a selection of multiwavelength all-sky images of the Galaxy.

The Galaxy is an Sbc/SBbc spiral galaxy and in the atlas it is classified in the normal (N) category of our sample, yet additional categories interacting (I), and active? (A?) (see Table 1, page 18) are also assigned. The Galaxy (along with the LMC and SMC) is classified I due to the existence of the H I Magellanic Stream (Wannier and Wrixon 1972; Murai and Fujimoto 1986) as shown in Figure 1.22. An A? classification is also given because the optical spectrum integrated over the central parsec of the Galaxy resembles a Seyfert galaxy (Mezger et al. 1996). This is consistent with results that suggest the presence of a several million $M_{\odot}$ black hole in the nucleus.

3.1 Quiet monster – Sagittarius A*

Our Galaxy's nucleus, some 8.5 kpc distant in the constellation of Sagittarius, has long been a target for multiwavelength observations. Because of the large amounts of obscuring dust towards the nucleus and in the disk of the Galaxy, infrared observations have proved invaluable in showing the content and structure of this region. Figure 3.1 shows a 48° by 33° IRAS IR image of the central region of the Galaxy. The disk of the Galaxy is the bright band running diagonally across the image.

The bright central region contains the nucleus. The concentrated blobs of yellow are giant clouds of interstellar gas and dust heated by recently formed stars and nearby massive, hot stars. The Galactic center harbors a strong radio continuum source, known as Sagittarius A or Sgr A, containing a bright source (Sgr A West)

within which there is a smaller, concentrated source called Sagittarius A* or Sgr A* (pronounced Sadge-A-star). The central part of the Galaxy has also been imaged at submillimeter wavelengths. A slice across the Galactic plane measuring 1450 light-years (~440 pc) is shown in Figure 3.2. This image is a mosaic of many smaller images taken over 15 nights (and spanning two years) using SCUBA at JCMT. The image shows emission at 850 μm and identifies the distribution of cold interstellar dust that resides in molecular clouds of gas.

Going in closer, Figure 3.3 shows a multiwavelength view in X-rays, optical and the infrared, using three of NASA's Great Observatories, Chandra, Hubble and Spitzer, respectively. Sgr A West is the strong emission region in white. VLA radio continuum observations of the central ~60 pc of the nucleus shows thin, collimated filaments that are probably produced by strong magnetic fields. These filaments run parallel for ~20 pc then bend downwards towards Sgr A West. Radio observations spanning only 7 pc (VLA 6 cm radio continuum; Figure 3.4) show the remarkable spiral-arm-like structure of Sgr A West. The emission is due to ionized gas that is being heated by numerous young, hot stars. At its center is Sgr A*, a point-like source that many suspect is the nucleus of the Galaxy.

Using the Berkeley Illinois Maryland Association (BIMA) array at Hat Creek, California, the molecular gas content near Sgr A* has also been observed. A ring of molecular hydrogen cyanide (HCN) at a temperature of ~300 K encircles the spiral-arm structure centered on Sgr A*. HCN is detected by observing the $(J = 1 \rightarrow 0)$ molecular transition lines at 86.34 and 88.62 GHz (0.35 and 0.34 cm, respectively). The molecular ring has a radius of ~2 pc and is inclined at ~70° to our line of sight.

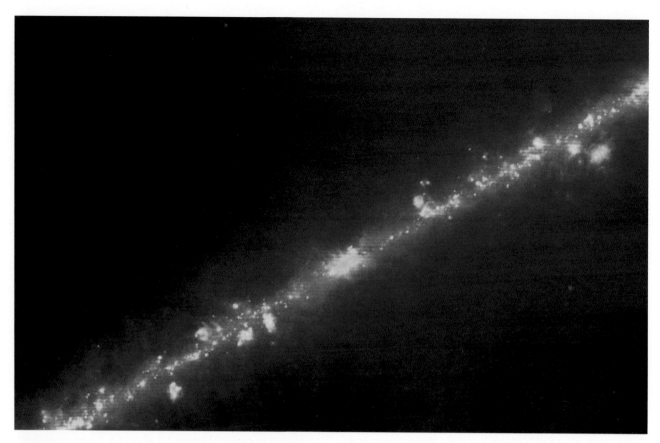

Figure 3.1

The disk of the Galaxy. The IRAS infrared image shows the central 48° by 33°
region of the Galaxy. The warmest dust is shown in blue whilst the coldest
is represented by red. The regions of yellow are giant clouds of interstellar
gas and dust.

Credit: Infrared Processing and Analysis Center, Caltech/JPL.

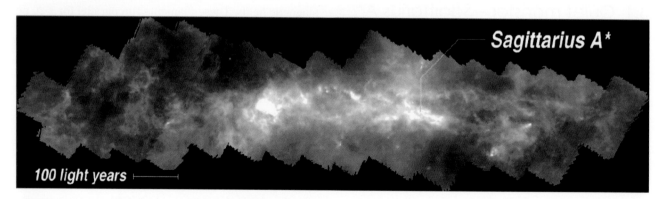

Figure 3.2

The Galaxy center. A JCMT SCUBA image at 850 μm spanning 1450 light-years
(∼440 pc). The position of Sgr A* is indicated. The image spans 2.9° in Galactic
longitude, and 0.8° in Galactic latitude.

Credit: D. Pierce-Price, University of Cambridge.

Figure 3.3

The Galaxy center by CXO, HST and Spitzer. A multiwavelength image spanning 32′ by 16′ with X-rays (blue, purple), optical (yellow) and IR (red) combined. The image spans ~60 pc (long axis) at the distance of the Galactic center.

Credit: X-ray: NASA/UMass/D. Wang et al. Optical: NASA/ESA/STScI/D. Wang et al. IR: NASA/JPL-Caltech/SSC/S. Stolovy.

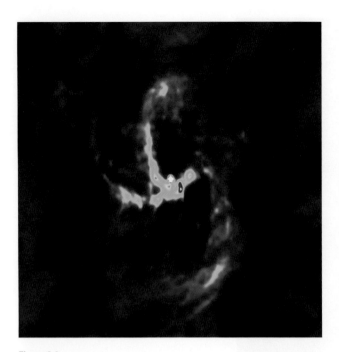

Figure 3.4

A VLA 6 cm radio continuum image of Sgr A West, showing a spiral-like pattern of thermal ionized gas. Its nucleus is Sgr A*. The image is 7 pc on a side.

Credit: NRAO/AUI/NSF.

Further observations reveal the unusual nature of the nucleus of the Galaxy. High-resolution imaging in the near-IR (Genzel *et al.* 1997) detects several fast-moving stars ($>10^3$ km s^{-1}) within 0.01 pc of Sgr A*. Based on the velocities, a mass of $\sim 3 \times 10^6$ M$_\odot$ was inferred within one light-week of the compact radio source. There is no stable configuration of normal stars, stellar remnants or substellar entities that can produce such a density and this strongly suggests the existence of a supermassive black hole at the center of the Galaxy.

ESO New Technology Telescope and VLT observations have now been made for over 10 years of a star, denoted S2, near Sgr A*. In 2002, S2 approached within 17 light-hours of Sgr A* with a velocity of 5000 km s^{-1}. Observations (Schödel *et al.* 2002) show that S2 is on a bound, highly elliptical, Keplerian orbit around Sgr A*, with a period of 15.2 years. The orbit requires a central point mass (Gillessen *et al.* 2009) of 4.3 ±0.5 × 10^6 M$_\odot$.

Between May and June, 2002, CXO observed X-ray outbursts from hot gas around Sgr A* on a daily basis. Surrounding Sgr A*, $\sim$20 million K gas was detected suggesting numerous outbursts have occurred over the previous 10,000 years in order to energize it. Archive data from the VLA has shown that Sgr A* emits radio emission

in a regular fashion, every 106 days. It is possible that gas in an accretion disk around the SMBH is being heated and forced outwards, emitting radiation in a periodic way.

3.2 All-sky maps of the Galaxy

Our Solar System is $\sim$8.5 kpc from the Galaxy nucleus, and is located in the Orion or Local Arm. The Galaxy is viewed from well inside its disk, which is 800–1000 pc thick at the location of the Sun. Each of the following images presents the whole sky in Galactic coordinates, with the plane of the Galaxy extending along the horizontal axis, and the nucleus of the Galaxy at the center. Each image (apart from Figure 3.9) is shown in Aithoff[1] projection. Galactic longitude (l) runs from 0° at the center to 180° at the left edge, and from 180° to 360° from the right edge back to the center. The north and south Galactic poles (latitudes $b = \pm90°$) are at the top and bottom, respectively, of each image.

3.2.1 Gamma ray – > 100 MeV

Figure 3.5 was taken by CGRO with EGRET. The image shows gamma-ray photons with energies greater than 100 MeV. The majority of these gamma rays are produced by collisions of cosmic rays with hydrogen in interstellar clouds. Hence the disk region is well defined; however the highest intensity sources are associated with pulsars and the Galactic center. The three bright sources near the plane at Galactic longitudes 185°, 195° and 265° (starting from the right edge and going left towards the center) are the Crab, Geminga and Vela Pulsars, respectively. The angular resolution of the image is $\sim$2°.

An all-sky view from Fermi and the LAT instrument sensitive from 30 MeV to greater than 300 GeV is shown in Figure 1.16.

3.2.2 Gamma ray – 1.8 MeV

Figure 3.6 was taken by CGRO with COMPTEL. The image shows gamma-ray photon events with energies of 1.8 MeV. This emission traces the decay of radioactive ^{26}Al which has a half-life of 7×10^5 yrs. ^{26}Al is believed to form in novae and supernovae, and in massive ($M > 10$ M$_\odot$) stars. The distribution of emission therefore traces recent nucleosynthesis sites.

---

1 Aithoff projection preserves the presentation of equal areas as would be seen on the surface of a sphere.

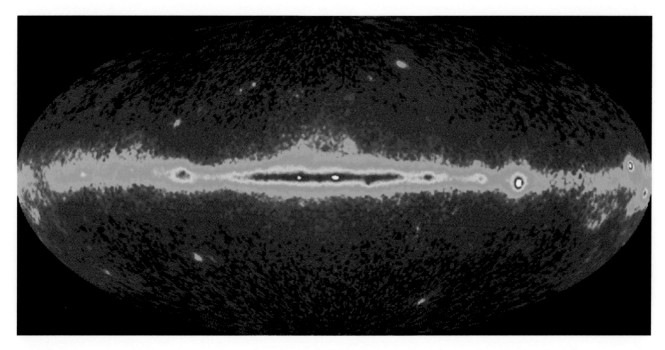

Figure 3.5

The Galaxy: gamma ray >100 MeV.

Credit: ADF/GSFC.

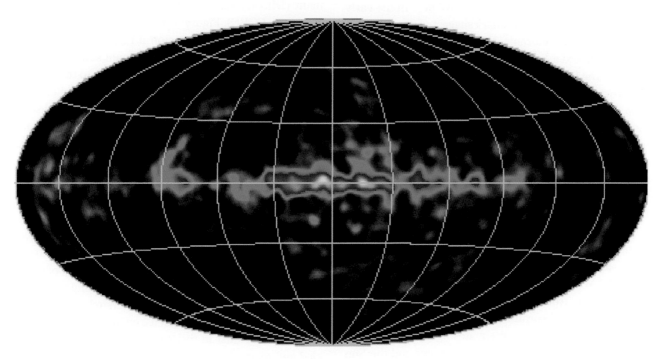

Figure 3.6

The Galaxy: gamma ray 1.8 MeV.

Credit: R. Diehl, U. Oberlack and the COMPTEL team.

3.2.3 Soft X-ray

Observed by the PSPC on board ROSAT, Figure 3.7 shows soft X-rays at energies of 0.25 keV (red), 0.75 keV (green) and 1.5 keV (blue). The reddest objects have the softest (lower) X-ray energies, whilst blue objects indicate harder (higher) energy sources. Soft X-rays are efficiently absorbed by the ISM, hence the observed colors can be influenced by the amount of absorption along our line of sight, and/or by the intrinsic temperature of each object. The point sources are binary star systems and pulsars (Vela is the bright source, center-right of the image). Extended emission is related to SNRs or hot gas outflows. The faint "spur" near the Galactic center and extending upwards is an outflow, and can also be seen as the North Polar Spur in the radio continuum (Figure 3.14) at 73 cm. Black regions are areas not observed by ROSAT. The angular resolution of the image is 115'.

3.2.4 Optical

A unique optical representation of the Galaxy has been produced at the Lund Observatory, Sweden. In the early 1950s, Professor Knut Lundmark suggested the construction of a Galaxy panorama. Martin Keskülä and Tatjana Keskülä painted a 1×2 m map showing 7000 individual stars and the faint, diffuse Galaxy as shown in Figure 3.8. The work took almost two years, and was completed in 1955. From the left, the constellations Auriga, Perseus, Cassiopeia and Cepheus are visible. The large dark parts of Cygnus and Aquila can be recognized to the left of center, where Sagittarius is found. Continuing right, Centaurus, Crux and Carina are seen. The fainter parts further to the right are Monoceros and Gemini, and farthest to the right is Auriga again.

The LMC and SMC are easily identified as discrete "blobs" well below the Galactic plane and to the right, and just below the plane, near the left edge, is NGC 224/M 31, the Andromeda Galaxy. The majority of emission is from nearby stars within the Orion Arm and 10^4 K gas. Absorption by interstellar dust at optical wavelengths strongly attenuates light. For example, the extinction between the Galactic center and the Sun corresponds to $A_V \sim 31$ magnitudes. Note that this is much greater than would be determined using the general relation given in Section 2.4.4 ($A_V = 13.6$ mag) highlighting the extreme absorption along this line of sight.

3.2.5 Optical – Hα at 6563 Å

Figure 3.9 shows optical line emission from Hα (6563 Å), ionized hydrogen gas. The image[2] has 6' (FWHM) resolution and is a composite of the Virginia Tech Spectral Line Survey (VTSS) in the northern hemisphere and the Southern H-Alpha Sky Survey Atlas (SHASSA) in the southern hemisphere. The Wisconsin H-Alpha Mapper (WHAM) survey is used to calibrate the data. VTSS used the Virginia Tech Spectral Line Imaging Camera (SLIC) at Martin Observatory in south-western Virginia. SHASSA observations were taken with a robotic camera at CTIO in Chile. WHAM used a 0.6 m telescope at KPNO in Arizona for two years from 1996. The surveys allow a study of the structure and kinematics of the warm ionized interstellar medium (WIM), a hydrogen gas plasma at 10^4 K. These images show "supershells" and "chimney" structures and allow a comparison with other phases of the ISM such as atomic hydrogen, H I (e.g. Reynolds *et al.* 1995).

3.2.6 Near-infrared

Observed by the Diffuse Infrared Background Experiment (DIRBE) instrument on the Cosmic Background Explorer (COBE), Figure 3.10 shows near-IR composite colors associated with emission at 1.25 μm (blue), 2.2 μm (green) and 3.5 μm (red). The dominant source of radiation is disk and bulge stars. In comparison to the optical, absorption is much reduced, and the structural characteristics of the Galaxy are clearly seen. This includes an increased stellar density towards the Galactic plane and the structure of the Galactic bulge and nucleus region. The angular resolution of the image is 42'.

3.2.7 Mid- and far-infrared

A color composite image of the mid- and far-infrared as observed by IRAS. Figure 3.11 shows 12 μm (blue), 60 μm (green) and 100 μm (red) emission. The majority of emission is thermal, as radiation from star-forming regions is absorbed and then re-emitted by interstellar dust. Zodiacal emission (previously discussed in Section 2.1) from interplanetary dust in the Solar System has been modeled and removed from the image. Black strips are gaps in the sky coverage of the survey. Bright star-formation regions are seen just above center (Ophiuchus) and at

---

2 Not in Aithoff projection – emission is plotted across the whole field of view.

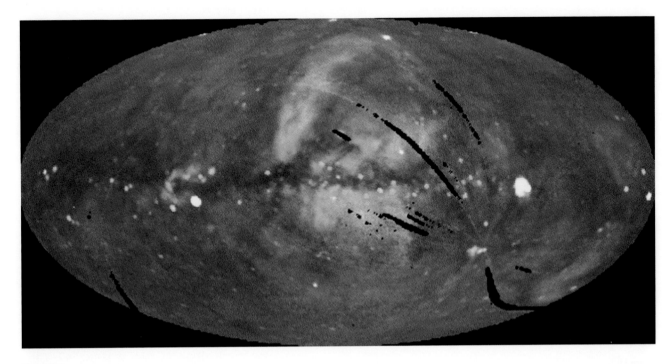

Figure 3.7

The Galaxy: soft X-ray.

Credit: ADF/GSFC.

Figure 3.8

The Galaxy: optical.

Credit: L. Lindegren, Lund Observatory.

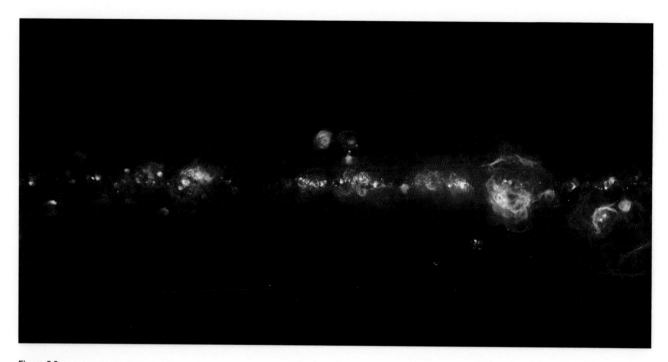

Figure 3.9

The Galaxy: optical – Hα at 6563 Å.

Credit: Permission granted by D. Finkbeiner. Original data from WHAM (funded primarily through the National Science Foundation, NSF), VTSS and SHASSA (both supported by the NSF). The SHASSA data is ©Las Cumbres Observatory, Inc.

Figure 3.10

The Galaxy: near-IR.

Credit: ADF/GSFC.

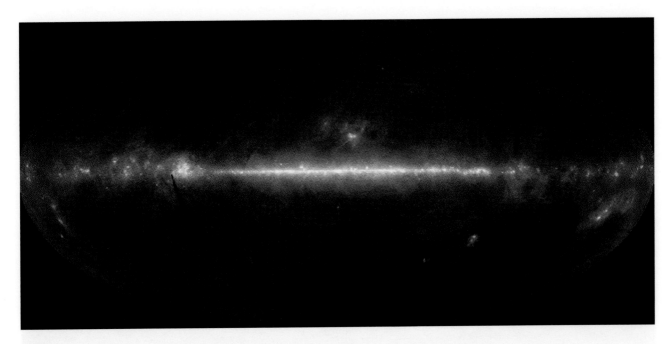

Figure 3.11

The Galaxy: mid- and far-IR.

Credit: ADF/GSFC.

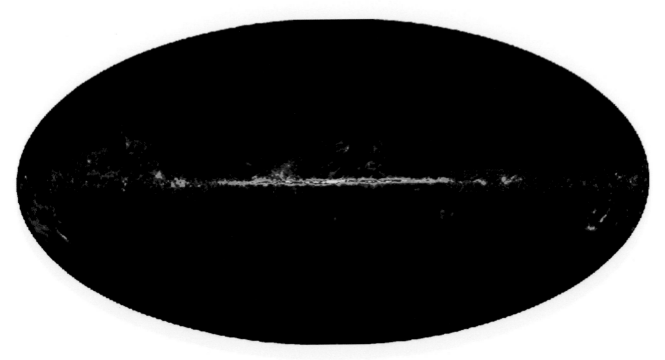

Figure 3.12

The Galaxy: radio H_2 (observed CO).

Credit: ADF/GSFC.

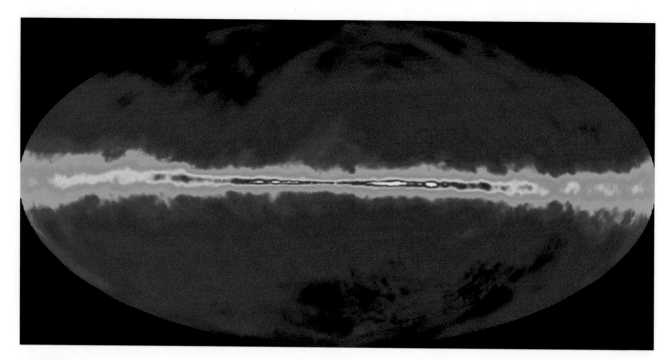

Figure 3.13

The Galaxy: radio H I.

Credit: ADF/GSFC.

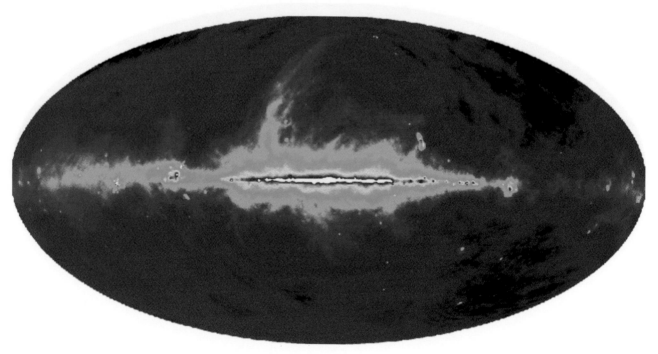

Figure 3.14

The Galaxy: radio 73 cm/408 MHz.

Credit: ADF/GSFC.

the far lower right (Orion). The image demonstrates how remarkably flat and thin the disk of the Galaxy is. The angular resolution of the image is $5'$.

3.2.8 Radio – molecular hydrogen

Figure 3.12 depicts the inferred molecular hydrogen (H_2) content from observations of the 2.6 mm CO ($J = 1 \rightarrow 0$) emission line. Such molecular gas is concentrated in the spiral arms and is associated with clouds of gas that are star-formation sites. Indirect methods must be used to determine the H_2 content since H_2 has few strong spectral signatures that are observable under typical interstellar conditions. Hence the 2.6 mm CO line is used as a tracer of the mass of H_2 (see Section 2.4.7). The angular resolution is $30'$, and the image shows column densities (N_{H_2}) of $12 \rightarrow 285 \times 10^{20}$ cm^{-2}.

3.2.9 Radio – atomic hydrogen

Figure 3.13 shows the atomic hydrogen (H I) content based on the 21 cm spectral line. The emission shows the cold ISM that predominantly lies close to the Galactic plane in a very thin layer, only $\sim$50–100 pc thick. The image is based mainly on the Leiden–Dwingeloo Survey of Galactic Neutral Hydrogen (Hartmann and Burton 1997). The survey took four years to complete using the Dwingeloo 25 m radio telescope. The angular resolution varies from $45'$ to $60'$, and the survey is sensitive to column densities (N_H) of $10 \rightarrow 230 \times 10^{20}$ cm^{-2}.

3.2.10 Radio – 73 cm/408 MHz continuum

Figure 3.14 shows the radio continuum at a wavelength of 73 cm or 408 MHz. It is derived from the all-sky data of Haslam *et al.* (1982) and combines data from Jodrell Bank, Effelsberg 100 m and Parkes 64 m telescopes. This image shows the distribution of relativistic electrons that are spiraling along magnetic field lines and emitting synchrotron radiation. The emission is intense along the plane of the Galaxy, with a maximum at the center, as well as "loops", "ridges" or "spurs" leaving the plane. The large spur, just left of center and curving upwards is the North Polar Spur, as also seen in Figure 3.7. High galactic latitudes show discrete radio sources. Centaurus A is seen at $l = 310°$ and $b = +20°$ (above the Galactic plane, middle right), whilst the LMC is at $l = 280°$ and $b = -31°$ (below the Galactic plane, middle right). The angular resolution of the image is $0.85°$.

3.3 Additional reading

The listing is not intended to be comprehensive but will complement the text.

A. Eckart, R. Schödel, C. M. Straubmeier, 2005, *The Black Hole at the Center of the Milky Way* (Imperial College Press, London).

PART 4

The Atlas

MULTIWAVELENGTH images of normal, interacting, merging, starburst and active galaxies are presented. Galaxies are ordered by right ascension in each category sample as listed in Table 1 (page 18). The images come from a variety of sources – some single-wavelength images are presented in false color, whilst multiwavelength images are color coded.

A brief description of each galaxy including its major multiwavelength properties (literature references start on page 243) is given. The description is ordered via wavelength, usually starting with X-ray properties and ending with radio properties. However, papers that present multiwavelength properties have their findings summarized concurrently. Each galaxy description is provided to highlight some of the more recent and interesting scientific findings. The description is, however, not intended to be a comprehensive description of the galaxy. Readers interested in comprehensive literature searches on individual galaxies should use a nearby mirror site of the NASA Astrophysics Data System listed in, e.g.

http://ads.nao.ac.jp/mirrors.html

or use the NASA/IPAC Extragalactic Database (NED)

http://nedwww.ipac.caltech.edu/

Technical information about the galaxy images – telescope/instrument, observer(s), λ/ν/energy/filter, exposure, resolution and image source – is listed in Appendix C (page 224). For each image north is up and east is to the left unless otherwise indicated. Inclinations of galaxies are referred to in the sense that $0°$ is face-on and $90°$ is edge-on to our line of sight. The spatial scale of each galaxy image is not always the same although the field of view is listed when it is known.

4.1 NORMAL GALAXIES

4.1.1 NGC 224/Messier 31

The SbI-II galaxy, NGC 224, Messier 31, is the nearest, large spiral galaxy, some 700 kpc distant. It is commonly called the Andromeda Galaxy. It is one of the best observed galaxies in the sky, and along with the Galaxy they comprise the two dominant galaxies in the Local Group. NGC 224/M 31 is ~2 times more luminous and more massive than the Galaxy. NGC 224/M 31 is approaching the Galaxy at ~300 km s^{-1} (heliocentric) and in several Gyr they will undergo a major merger event. NGC 224/M 31 has more than 15 dwarf satellite galaxies and this number will undoubtedly increase with current and future deep imaging surveys of the region (i.e. McConnachie *et al.* 2009). It is inclined at 12.5° to our line of sight.

A multicolor ultraviolet and infrared image of NGC 224/M 31 is shown in Figure 1.4.

An optical stellar density map of the Andromeda–Triangulum region with scale images of the disks of NGC 224/M 31 and NGC 598/M 33 is shown in Figure 2.16.

■ NGC 224/Messier 31:

A ROSAT PSPC survey (Supper *et al.* 1997) detected 396 individual sources of which 327 were not associated with sources discovered from the previous *Einstein* Observatory survey. Twenty-nine sources are identified with globular clusters and 17 with SNRs. Supper *et al.* (2001) reports on a second survey with ROSAT PSPC in 1992. Comparing the two ROSAT surveys, 34 long-term variable sources and eight transient candidates are found. The observed luminosities range from 4×10^{35} to 4×10^{38} erg s^{-1}. Kong *et al.* (2002) observed the central region with CXO ACIS and detected 204 X-ray sources with a detection limit of $\sim 2 \times 10^{35}$ ergs s^{-1}. Of these, 22 are identified with globular clusters, two with SNRs, nine with planetary nebula, and nine with supersoft sources. About 50% of the sources are variable on time-scales of months.

The HST FOC far-UV image (King *et al.* 1992; King, Stanford and Crane 1995; Maoz *et al.* 1996) of the nucleus shows bright, centrally peaked, asymmetric emission. King *et al.* (1995) show this peak is cospatial with the fainter of two peaks in the optical WFPC images (Lauer *et al.* 1993) that is the dynamical center. Bender *et al.* (2005) identify a (third) blue nucleus, P3, in the lower brightness P2 peak (Figure 4.4) and confirm earlier predictions that P3 harbors a SMBH, of 2×10^8 M$_\odot$.

Ibata *et al.* (2001) report an excess or "stream" of metal-rich RGB stars in the halo that could be due to the close companion dwarfs NGC 221/M 32 and NGC 205/M 110 losing stars due to tidal interactions. See also the optical stellar density map of the Andromeda–Triangulum region in Figure 2.16.

Haas *et al.* (1998) present a 175 μm map that shows cold dust, dominated by a ring at 10 kpc radius supplemented by a faint outer ring at 14 kpc. No clear spiral pattern is recognizable.

Nieten *et al.* (2006) use IRAM to study the CO ($J = 1 \rightarrow 0$) distribution of molecular gas. CO is concentrated in narrow filaments often coinciding with dust lanes. Between 4 and 12 kpc radius the brightest CO filaments define a two-armed spiral pattern.

Berkhuijsen, Beck and Hoernes (2003) present maps of thermal and non-thermal emission at 6.2 cm showing that stronger total emission in the north is related to stronger thermal emission. Minor axis profiles of non-thermal and polarized emission are nearly identical – this suggests that recent star formation does not lead to a local increase of the magnetic field strength. On scales of 600 pc, dense H II regions do not cause significant Faraday rotation or depolarization, hence, rotation measures and Faraday depolarization must originate in the diffuse ionized gas.

The high angular resolution WSRT H I image of Brinks and Shane (1984) shows the cold gas to be distributed in a ring 10 kpc from the nucleus. The complete H I distribution extends to ∼40 kpc from the center and outlines the general spiral arm structure. Small holes or voids in the cold gas distribution are probably caused by intense star formation.

The radio continuum at 1.412 GHz using WSRT has been mapped by Walterbos, Brinks and Shane (1985) and is dominated by two emission regions: the nucleus, and a ring ∼10 kpc from the nucleus.

A monograph review of NGC 224/M 31 by Hodge (1992) is a valuable overview.

Figure 4.1

NGC 224/M 31: X-ray (0.3–12.0 keV), 15′ across. Variable sources indicated.

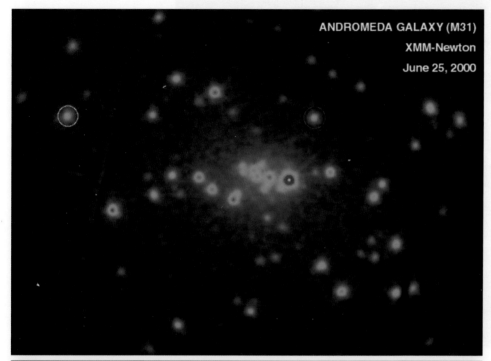

ANDROMEDA GALAXY (M31)
XMM-Newton
June 25, 2000

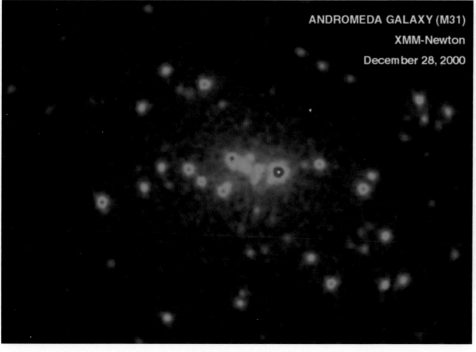

ANDROMEDA GALAXY (M31)
XMM-Newton
December 28, 2000

Figure 4.2

NGC 224/M 31: X-ray (0.1–1.0 keV, red; 1.0–2.0 keV, green; 2.0–4.0 keV, blue), 28′ across.

Figure 4.3

NGC 224/M 31: Far-UV + mid-UV, 1900–2600 Å, 100′ across, N is 55° counter-clockwise (CCW) from up.

Figure 4.4

NGC 224/M 31: Optical – nucleus, F160BW, F300W, F555W, F814W, 11.6′ across, N is 55° CCW from up. The nuclear brightness peaks are P1 (left), P2 (right).

Figure 4.5

NGC 224/M 31: Optical B.

Figure 4.6

NGC 224/M 31: Near-IR 1.2 μm (blue), 1.6 μm (green), 2.2 μm (red), satellite galaxies NGC 221/M 32 is lower, middle; NGC 205/M 110 is upper, right, 1° across.

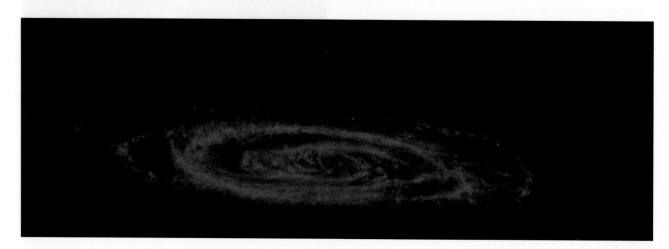

Figure 4.7

NGC 224/M 31: Mid-IR 8.0 μm, 3.1° across, N is 55° CCW from up.

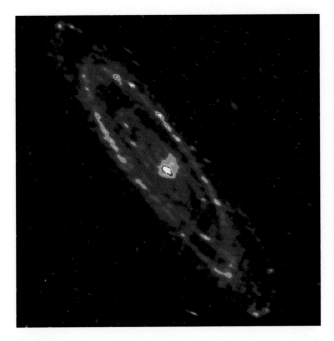

Figure 4.8

NGC 224/M 31: Far-IR 60 μm, 2° across.

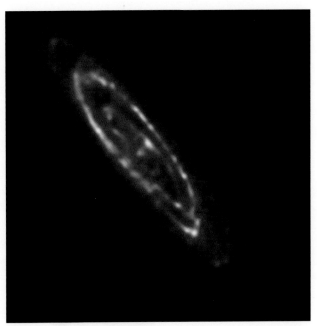

Figure 4.9

NGC 224/M 31: Far-IR 175 μm.

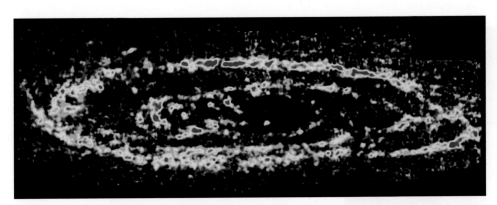

Figure 4.10

NGC 224/M 31: Radio CO 2.6 mm. N is 55° CCW from up. Copyright: ©MPIfR Bonn/IRAM (Ch. Nieten, N. Neininger, M. Guelin *et al.*).

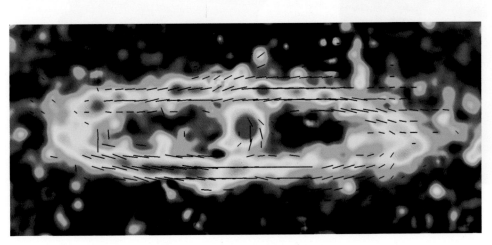

Figure 4.11

NGC 224/M 31: Radio 6 cm continuum, and magnetic field (vectors). N is 55° CCW from up. Copyright: ©MPIfR Bonn (R. Beck, E.M. Berkhuijsen, P. Hoernes).

Figure 4.12

NGC 224/M 31: Radio combined VLA and Effelsberg 20 cm.

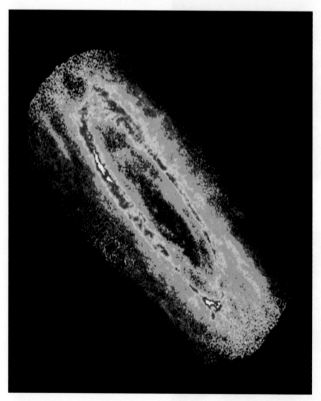

Figure 4.13

NGC 224/M 31: Radio H I – intensity.

Figure 4.14

NGC 224/M 31: Radio H I – velocity.

4.1.2 Small Magellanic Cloud

The Small Magellanic Cloud, an ImIV-V galaxy, is a close neighbor of the Galaxy, being only 70 kpc away (not much more than two Galaxy diameters). It, along with the Large Magellanic Cloud, is named after the Portuguese explorer Ferdinand Magellan who captained a voyage around the world in 1518–1520. He failed to complete the voyage, however, the Clouds were mentioned in the ships' journals as they sailed the South Pacific and his name later became associated with them. Both the SMC and LMC are easily visible naked eye objects to observers at latitudes below $-30°$. The SMC subtends about $4°$ of the sky.

A multicolor optical image of the SMC is shown in Figure 1.2.

Figure 1.22 shows the Magellanic Stream in relation to the SMC.

A radio image showing the H I distribution of the SMC is shown in Figure 2.15.

Due to the presence of the Magellanic Stream a secondary classification of I is adopted.

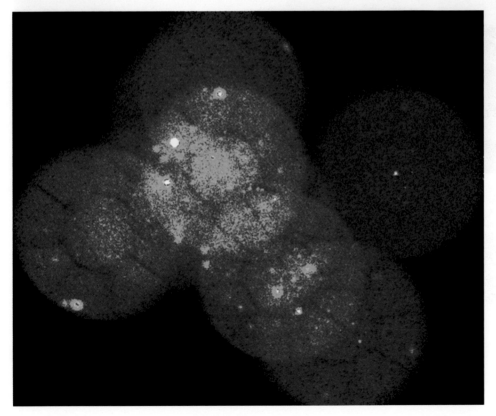

Figure 4.15

SMC: Soft X-ray (0.1–2.4 keV), 5.5° across.

Figure 4.16

SMC: Near-IR 3.6 μm (blue), mid-IR 8.0 μm (green), mid-IR+far-IR 24, 70 and
160 μm (red), ~5.4° across. The stellar cluster (blue) to the N is the Galactic
globular cluster NGC 362. Emission in green (left, top and bottom) is from dust in
the Galaxy. N is 20° CW from up.

■ SMC:

Kahabka and Pietsch (1996) used nine ROSAT PSPC observations over a 2 yr period to search for X-ray binary systems and detected seven spectrally hard sources, and four spectrally soft sources. Kahabka *et al.* (1999) present results from a systematic search for point-like and moderately extended soft (0.1–2.4 keV) X-ray sources with ROSAT PSPC between October, 1991 and October, 1993. They detect 248 objects which are included in the first version of a SMC catalogue of soft X-ray sources.

Using HST/ACS two-filter photometry Sabbi *et al.* (2009) study the stellar formation history and find a first epoch ~12 Gyr ago – with star formation up to 2–3 Gyr ago, then an increase in star formation in the bar and wing ~500 Myr ago.

Staveley-Smith *et al.* (1997) present an ATCA survey of H I (Figure 2.15) that shows the cold gas distribution is complex, but can be understood in terms of numerous shells and supershells (radii > 300 pc)

driven by supernovae and high-mass stellar winds. The mean age of star formation, based on kinematical arguments, is 5.4 Myr. Stanimirovic *et al.* (1999) present Parkes 64 m and ATCA H I observations that are sensitive to spatial scales between 30 pc and 4 kpc which test the H I spatial power spectrum. The kinematics of the H I reveal three supergiant shells which were previously undetectable in the ATCA data alone. These shells have diameters up to 1.8 kpc and require energies up to 2×10^{54} erg for their formation.

The Parkes 64 m telescope observed that the radio continuum structure of the SMC (Haynes *et al.* 1986, 1991) is described by thermal emission associated with H II regions, and an extended non-thermal component that encompasses the whole galaxy. Filipovic *et al.* (2002) present continuum images using the Australia Telescope Compact Array (ATCA) and the Parkes 64 m and compile a new catalogue of sources at 1.42, 2.37, 4.80 and 8.64 GHz.

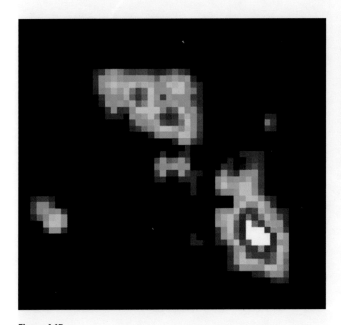

Figure 4.17
SMC: Radio CO.

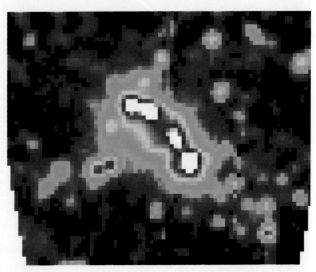

Figure 4.18
SMC: Radio 1.4 GHz continuum.

4.1.3 NGC 300

NGC 300 is an ScII.8 galaxy in the South Polar Group (or Sculptor Group) at a distance of 1.2 Mpc. The galaxy is inclined at ~40° and possesses a compact, bright nucleus and closely wound spiral arms.

Figure 4.19

NGC 300: X-ray 0.3–1.0 keV (red), 1.0–2.0 keV (green), 2.0–10.0 keV (blue). The white line indicates the optical extent.

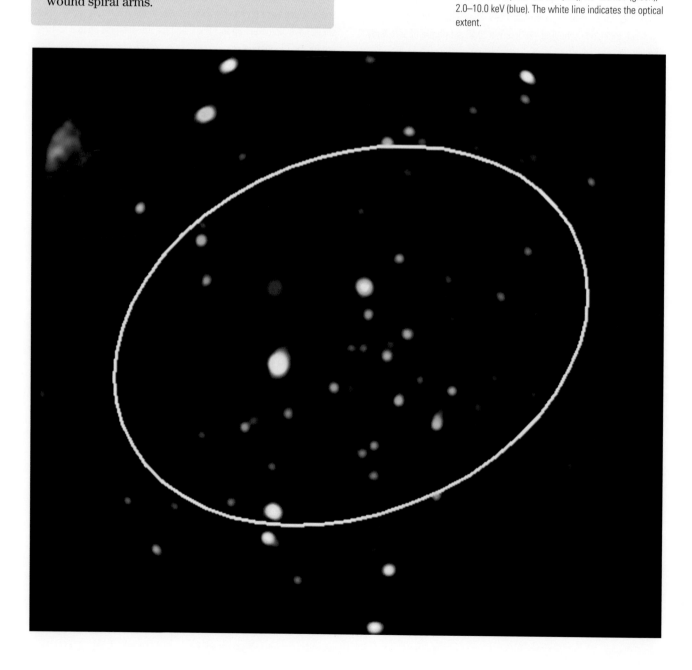

NGC 300:

Read and Pietsch (2001) report on ROSAT PSPC and HRI observations. A black hole X-ray binary candidate, a supersoft source and several SNRs and H II regions are detected. Unresolved emission comprises ~20% of the total X-ray (0.1–2.4 keV) luminosity of 5.8×10^{38} erg s^{-1}. Carpano *et al.* (2007) discovered NGC 300 X-1, a Wolf–Rayet/compact X-ray binary system, with a period of 32.8 h.

The HST FOC mid-UV image (Maoz *et al.* 1996) shows many point-like sources, with the brightest at the nucleus which is resolved (FWHM ~0.2″).

Davidge (1998) studied the evolved stellar content with deep J, H and K imaging noting significant numbers of ~10 Gyr old AGB stars suggesting that the disk contains an underlying old population.

The IRAS 60 μm emission (Rice 1993) is dominated by large, star-forming regions.

Puche, Carignan and Bosma (1990) find that the VLA observed H I distribution is very extended, and warped at large radii. A global mass-to-light ratio of 11 M$_\odot$/L$_{\odot B}$ is found, suggesting that dark matter in the South Polar Group is concentrated around the member galaxies. Condon (1987) lists NGC 300 as a possible VLA continuum detection at 1.49 GHz.

Figure 4.20

NGC 300: Far-UV 1400–1700 Å (blue), optical (red, yellow), 25′ across.

Figure 4.21

NGC 300: Optical, F450W (blue), F555W (green), F814W (red) – 1.8′ across, offset from nucleus, N is ~40° CCW from up.

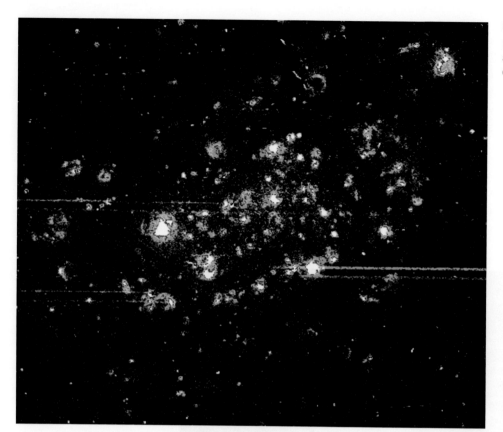

Figure 4.22

NGC 300: Optical Hα + [N II], 15′ across. (The lines are CCD readout defects.)

Figure 4.23

NGC 300: Near-IR 3.6 μm (blue), 4.5 μm (green), 5.8 μm (orange), mid-IR 8.0 μm (red), N is ~30° CW from up.

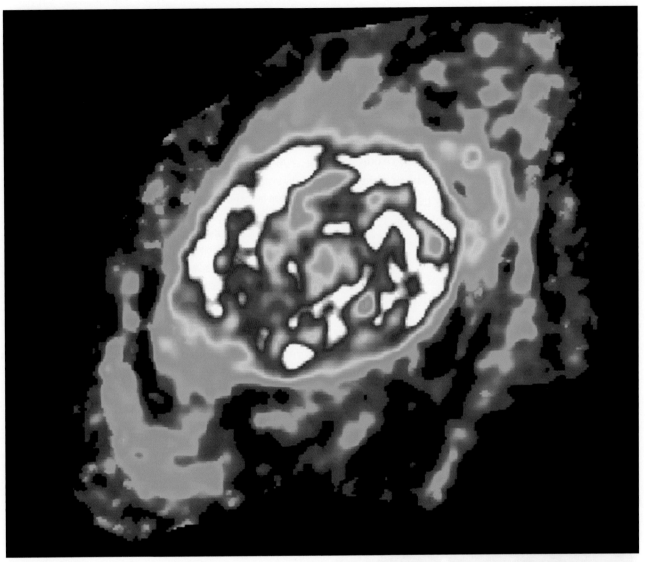

Figure 4.24

NGC 300: Radio H I, 40′ across.

4.1.4 NGC 598/Messier 33

Well defined, open spiral arms characterize NGC 598/Messier 33, an Sc(s)II-III Local Group member. It is inclined at 57° to the line of sight and is 700 kpc distant. The galaxy contains the bright H II region, NGC 604, NE of the nucleus.

A multicolor optical image of NGC 598/M 33 is shown in Figure 1.9.

An optical stellar density map of the Andromeda–Triangulum region with scale images of the disks of NGC 224/M 31 and NGC 598/M 33 is shown in Figure 2.16.

■ NGC 598/M 33:

A ROSAT PSPC study (Long *et al.* 1996) found the majority of sources are compact X-ray binaries. Approximately 31% of the non-nuclear soft X-ray flux is from diffuse emission. The nuclear X-ray source, M 33 X-8, has $L_X \sim 10^{39}$ erg s^{-1}, low for an AGN, but high for a Galactic source (e.g. a binary system). M 33 X-8 shows evidence of variability with a brightness change of factor ~ 2 over six months, suggesting that it is not a chance superposition of two or more binary systems. Orosz *et al.* (2007) derive a mass of 15.7 $M_\odot$ for the black hole in the system M 33 X-7, an eclipsing binary with a 3.5 day orbit around its ~ 70 $M_\odot$ companion.

Rice *et al.* (1990) present IRAS maps that show intense IR emission cospatial with the nucleus and bright H II regions. A CO study by Wilson and Scoville (1990) suggests that the NGC 598/M 33 value of α (the conversion factor from CO flux to H_2 column density, see Section 2.4.7) is very similar to that in the Galaxy. Rosolowsky *et al.* (2007) present high-resolution, 13″, CO maps of the central 5.5 kpc. Newton (1980) presented H I observations showing spiral structure well outside the brightest optical extent.

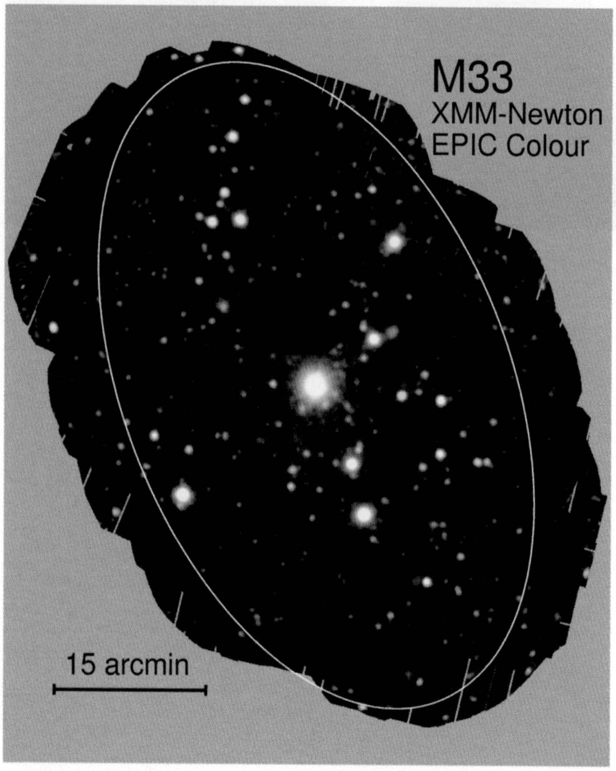

Figure 4.25

NGC 598/M 33: X-ray: 0.2–1.0 keV (red), 1.0–2.0 keV (green), 2.0–12.0 keV (blue).

M33
XMM-Newton
EPIC Colour

15 arcmin

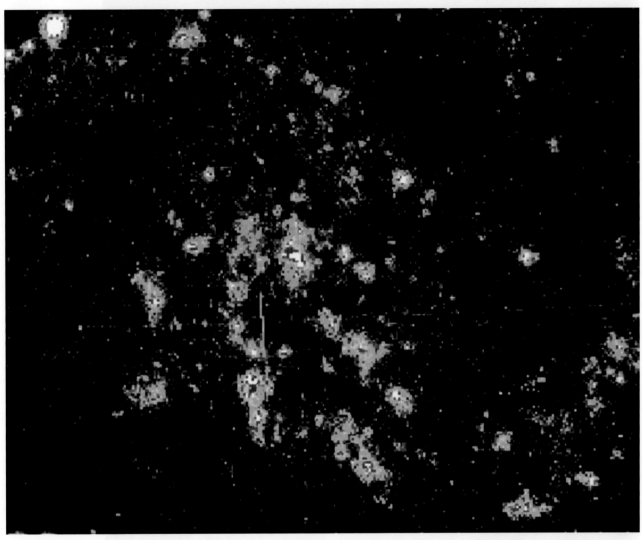

Figure 4.26
NGC 598/M 33: Far-UV 1600 Å, 25′ across.

Figure 4.27

NGC 598/M 33:
Far-UV (blue),
near-UV (green),
near-IR
(yellow/orange),
far-IR (red), 55′
across.

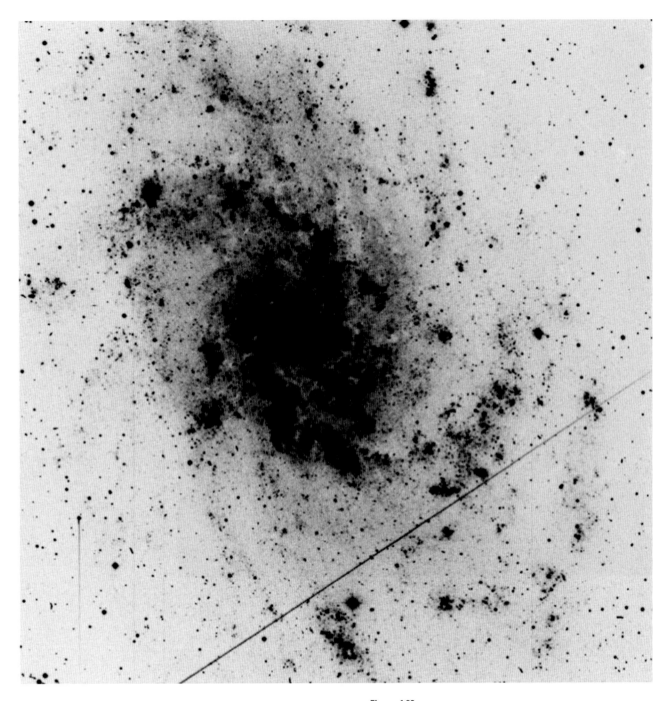

Figure 4.28

NGC 598/M 33: Optical U, large H II region NGC 604 is positioned left, top in spiral arm. Satellite trail to south and west.

Figure 4.29

NGC 598/M 33: Optical – H II region NGC 604, F336W (U), F375N ([O II]), F487N
(Hβ), F502N ([O III]), F555W (V), F656N (Hα), F658N ([N II]), F673N ([S II]), F814W
(I), F953N ([S III]);
2′ across, N is 10° CW from up.

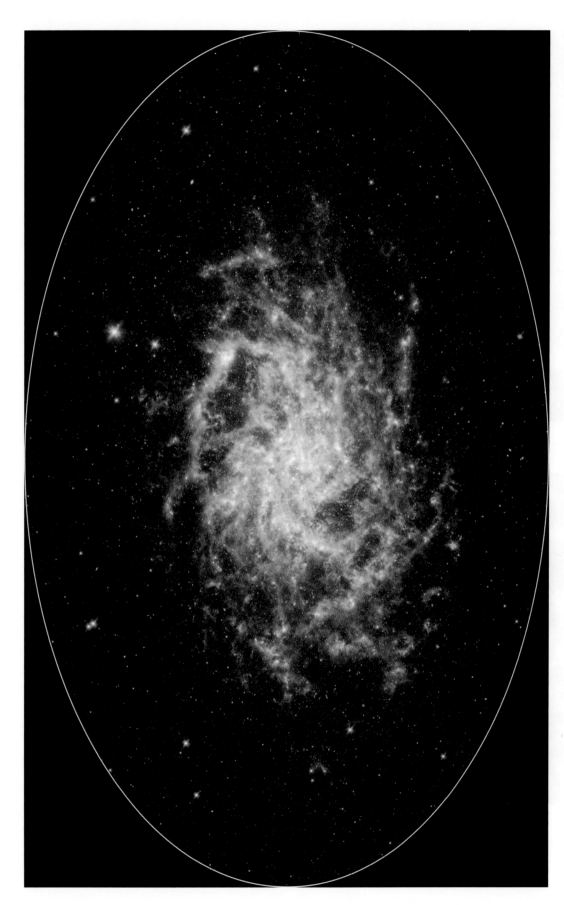

Figure 4.30

NGC 598/M 33:
Near-IR 3.6, 4.5 μm
(blue), mid-IR 8 μm
(green), 24 μm (red).
N is 5° CW from up.

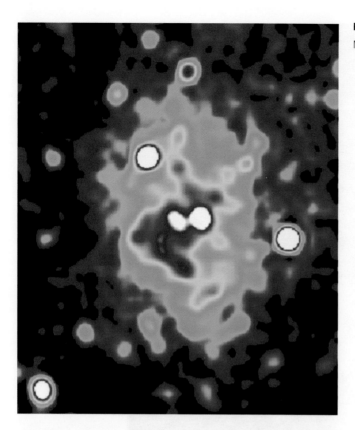

Figure 4.31
NGC 598/M 33: Radio 6 cm continuum.

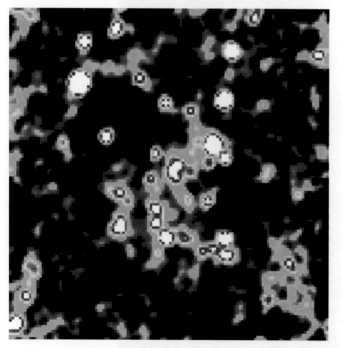

Figure 4.32
NGC 598/M 33: Radio 1.49 GHz continuum, 26 ′ across.

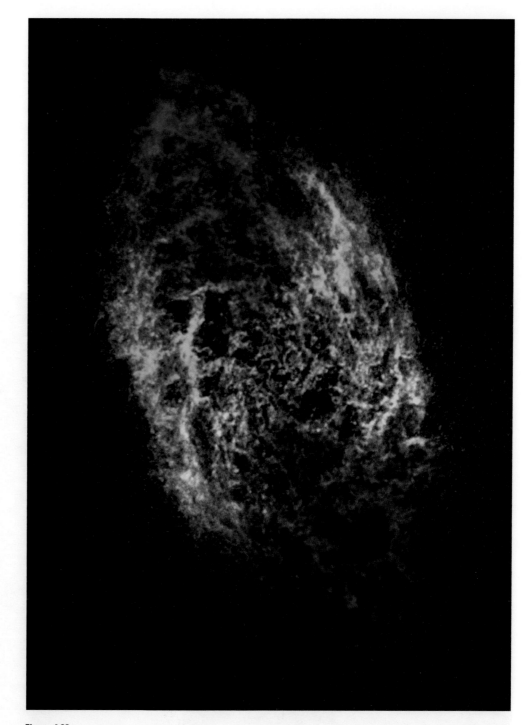

Figure 4.33

NGC 598/M 33: Radio H I – velocity.
Doppler redshifts and blueshifts are
relative to the center of mass. Brightness
is proportional to H I column density.

4.1.5 NGC 891

NGC 891 is an Sb edge-on galaxy, 9.6 Mpc distant in the NGC 1023 group. It is useful for studies of the minor-axis and above-plane of disk (vertical) distribution of stars and gas in a spiral galaxy.

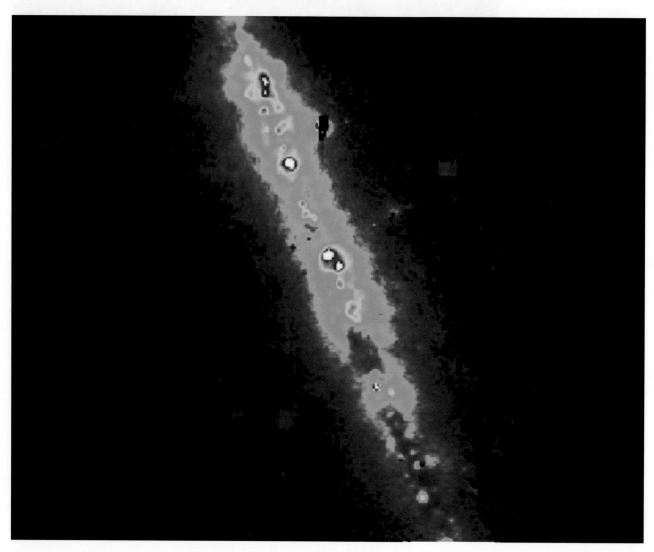

Figure 4.34

NGC 891: Optical Hα – central.

■ NGC 891:

Rand, Kulkarni and Hester (1990) studied the warm, 10^4 K ionized medium (WIM) using the Palomar 1.5 m telescope. An observed high level of star formation probably leads to NGC 891 having a larger surface density of diffuse ionized gas and a thicker ionized gas layer than the Galaxy. Vertical Hα filaments, extending more than 2 kpc from the plane, are seen.

These filaments have been interpreted as a result of correlated SNe explosions that produce expanding H I superbubbles that penetrate the disk. Hot, ionized gas can then escape the disk through these voids.

Shell structures at smaller vertical heights above the disk are also seen, and are energized by nearby large, active regions of star formation (OB associations). Howk and Savage (2000) present deep B, V, I and Hα images providing a view of two physically distinct "phases" of the thick interstellar disk. A dense and likely cold phase of the thick disk ISM (observed via B, V and I) and a warm ionized phase of the ISM away from the galactic plane (observed via Hα) exist.

Oosterloo, Fraternali and Sancisi (2007) present deep H I observations that show a large cold gas halo containing 30% of the total H I. A filament extends 22 kpc above the disk.

Sciama (1993) extensively discusses the origin of diffuse ionized gas in this galaxy. He argues that the ionization source could be decaying neutrinos with a mass ∼30 eV and a lifetime of ∼2×10^{23} sec.

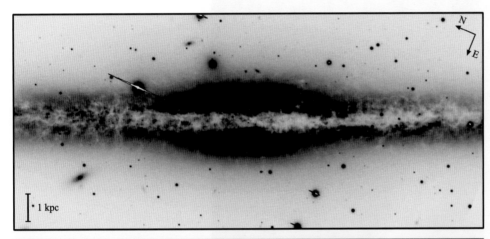

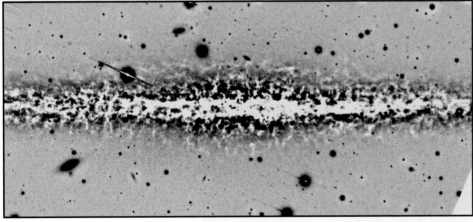

Figure 4.35

NGC 891: Optical V (top) and unsharp masked (bottom). Copyright WIYN Consortium Inc., all rights reserved.

Figure 4.36
NGC 891: Near-IR 1.2 μm (blue), 1.6 μm (green), 2.2 μm (red), 7′ across.

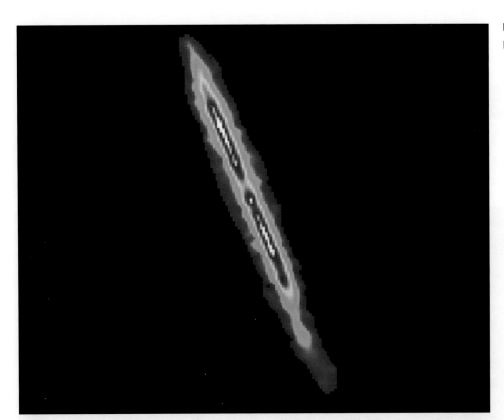

Figure 4.37
NGC 891: Radio H I, 12′ vertically.

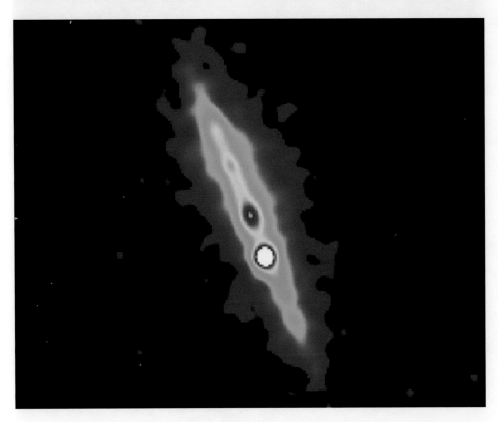

Figure 4.38
NGC 891: Radio continuum near H I.

4.1.6 NGC 1399

NGC 1399 is the central, dominant E1 galaxy in the Fornax cluster, 16.9 Mpc distant.

Figure 4.39

NGC 1399: X-ray, 3′ across.

■ NGC 1399:

ROSAT PSPC measurements (Rangarajan *et al.* 1995) indicate substantial amounts ($\sim$2 M$_\odot$ yr^{-1}) of cooling, 10^7 K gas exist in the center of the galaxy. Jones *et al.* (1997) use ROSAT PSPC to derive a hot gas temperature of 1.3 keV and a heavy element abundance of 0.6 of solar. Irwin *et al.* (2010) use CXO ACIS to discover an ultraluminous X-ray (ULX) source in a globular cluster. The authors suggest the ULX to be a >100 M$_\odot$ intermediate-mass black hole (IMBH[1]). Optical spectra of the object reveal emission from gas rich in oxygen and nitrogen suggesting that a white dwarf has been torn apart by the IMBH.

Goudfrooij *et al.* (1994b) detect nuclear Hα+[N II] emission. Lauer *et al.* (1995) use HST PC to describe the surface brightness profiles in the inner few parsecs of a sample of early-type galaxies that includes NGC 1399. They find that the inner profile exhibits a core or shallow cusp to within 2 pc ($\sim$0.02") of the nucleus. Grillmair *et al.* (1999) use WFPC2 data to study the luminosity and B − I color distribution of globular clusters. The color distributions of clusters in the central region of NGC 1399 and its nearby neighbor NGC 1404 are bimodal. The metallicity spread (inferred from the color distribution) is very similar to that of the giant elliptical NGC 4486/M 87 in the Virgo Cluster.

The double lobed, low-luminosity ($\sim$10^{39} ergs s^{-1}) radio source PKS 0336 − 355 (VLA radio continuum) is contained well within the optical galaxy and is probably thermally confined by the hot ISM (Killeen, Bicknell and Ekers 1988).

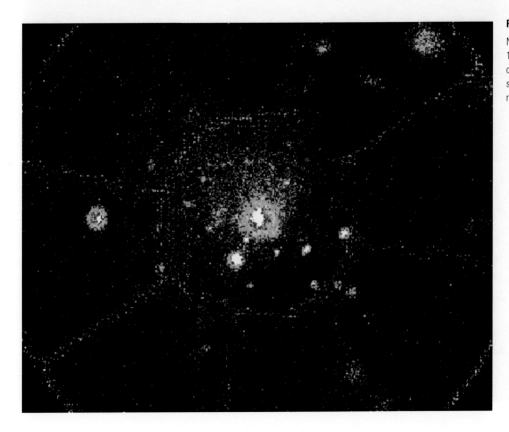

Figure 4.40

NGC 1399: Soft X-ray (0.1–2.4 keV), 1.7° across. NGC 1399 centered; other Fornax cluster galaxies are shown. ROSAT PSPC window support rib and ring structures visible.

---

1 Given some assumptions ULX sources exceeding 10^{39} erg s^{-1} are suggestive of the presence of an accreting black hole greater than 10^2 M$_\odot$. The inference is also model dependent and whilst numerous objects have now been found with these properties the existence of IMBHs is in dispute.

Figure 4.41

NGC 1399: Optical, 3′ across, N is 20 degrees
CCW from up.

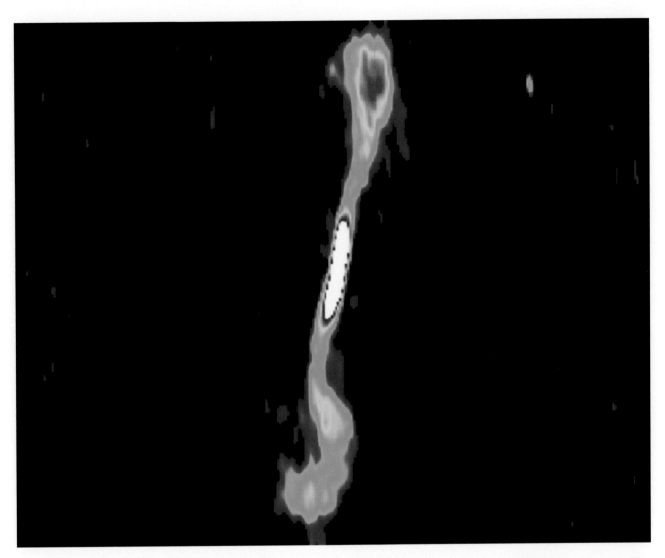

Figure 4.42

NGC 1399: Radio 4.9 GHz continuum, 4′ vertically.

4.1.7 Large Magellanic Cloud

The nearby Large Magellanic Cloud (LMC) at a distance of 50 kpc is slightly closer than the SMC. It has long been a laboratory for the study of star formation and is well studied at many wavelengths. Subtending about 7° of the sky along its long axis it is approximately 10% of the mass of our Galaxy. It is type SBmIII, with a prominent diffuse stellar bar. It has probably undergone a gravitational interaction with the Galaxy, as inferred by the presence of the H I Magellanic Stream. The ages of globular clusters in the LMC range from ~12 Gyr to ~10 Myr and provide a way of age-dating the star-formation epochs in the galaxy. The general stellar population has been summarized previously in Section 1.7.

The Tarantula Nebula, also known as 30 Doradus (30 Dor) and NGC 2070, dominates one end of the stellar bar and is a large H II region about 900 light-years in diameter. 30 Dor acquired its name because the ionizing source of the region was thought to be a single, massive star. This region has now been resolved into a group (R136) of many luminous stars. The filamentary and shell-like structure of the nebula is created by supernova explosions.

In February, 1987, supernova 1987A was detected in the LMC and near to 30 Dor. It was the nearest observed supernova since the era of Kepler. SN 1987A, a peculiar type II SN, provided (and continues to provide) a wealth of information about supernova explosions and the early development of a SNR. Interestingly the progenitor star, Sanduleak −69 202 was a blue supergiant (B3 I). Up until SN 1987A, it was thought that type II SNe progenitors were red supergiants.

Figure 1.22 shows the Magellanic Stream in relation to the LMC.

A multicolor optical image of the LMC in shown in Figure 1.24.

An H I radio image of the LMC in shown in Figure 1.25.

An image of the LMC in Hα is shown in Figure 2.9.

A secondary classification of I is adopted.

■ LMC:

Porter *et al.* (2009) have resolved the gamma-ray emission from the LMC. The LMC is observed between 200 MeV and 100 GeV and the signal is dominated by emission from the star-forming region 30 Doradus. The overall gamma-ray emission does not seem to correlate with the molecular gas distribution but better matches the atomic H I distribution.

Snowden and Petre (1994) present a mosaic of X-ray ROSAT PSPC observations. Preliminary analysis suggests that twice as many discrete sources are found in areas where the *Einstein* IPC surveyed.

Harris and Zaritsky (2009) present wide-field UBVI photometry of 20 million stars to produce a spatially resolved star-formation history.

After an initial burst, a quiescent period occurred between 12 and 5 Gyr ago, followed by star-formation peaks 2 Gyr, 500, 100 and 12 Myr ago. This history matches many similar periods in the SMC stellar history.

Cohen *et al.* (1988) surveyed the central $6° \times 6°$ in the $(J = 1 \to 0)$ transition of CO with the Columbia 1.2 m Millimeter-Wave Telescope. Emission appears over a large fraction of the LMC area and is dominated by a large complex near the giant H II region 30 Dor. The complex extends southwards with a high-velocity component. This gas, which could be part of a ring or shell, is probably a remnant from the progenitor gas that 30 Dor formed from, but is now expanding due to stellar winds from massive stars or supernovae.

Fukui *et al.* (2008) present a molecular cloud survey using NANTEN. Physical properties were derived for 164 giant molecular clouds (GMCs) out of a total 272 detected GMCs. Ott *et al.* (2008) study the molecular gas (CO, HCO^+ and HCN) along a 2 kpc ridge south of 30 Dor. CO and H I do not necessarily coincide and dense HCO^+ and HCN is related to star formation along filaments and shells due to nearby massive stars.

Haynes *et al.* (1986, 1991) present Parkes 64 m radio continuum maps that show emission closely matching emission from young Population I sources (e.g. Hα) which is of thermal origin. Emission from the 30 Dor complex dominates the images. Extended emission over the body of the

galaxy, which is of non-thermal origin, is consistent with the existence of strong, structured magnetic fields. The overall radio emission and magnetic fields may, however, be strongly influenced by ram pressure effects as the LMC passes through the halo of the Galaxy.

Kim *et al.* (1998) present an H I aperture synthesis mosaic (Figure 1.25) from 1344 separate pointings using the ATCA. The mosaic emphasizes the turbulent and fractal structure of the ISM. The structure of the neutral atomic ISM in the LMC is dominated by H I filaments, shells and voids. The bulk of the H I resides in a disk 7.3 kpc in diameter.

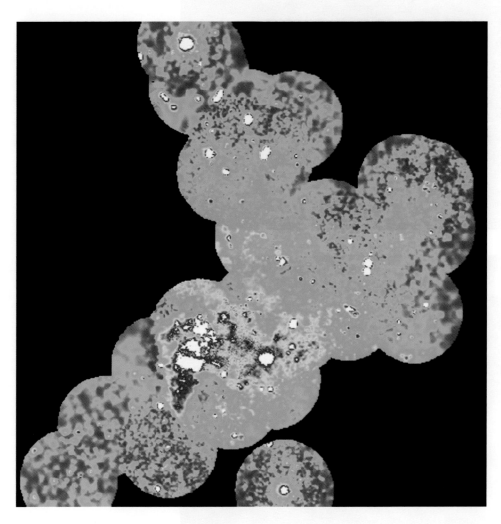

Figure 4.43

LMC: Soft X-ray (0.5–2.0 keV), RA spans 4 h 32 m (right, bottom) to 6 h 24 m (left, bottom). Dec. spans −63° 36′ (top) to −72° 12′ (bottom). The major emission region shown is the 30 Doradus/Tarantula Nebula.

Figure 4.44

LMC: Optical [O III] (blue), [S II] (red), Hα (green) – NGC 2074 (central) and environment, 38′ across. Credit: U. Michigan/CTIO MCELS Project/NOAO/AURA/NSF.

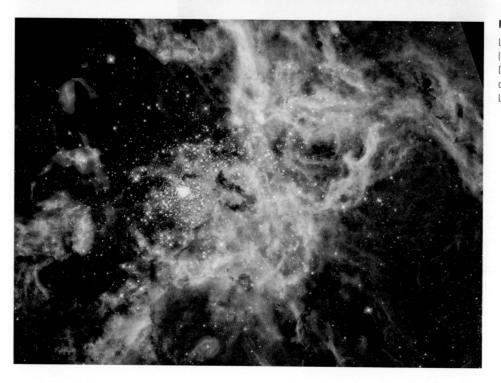

Figure 4.45

LMC: Optical F336W, F555W, F656N (Hα), F673N ([S II]), F814W – 30 Doradus (star cluster R136, left, center), 200 light-years across (at LMC), N is 60° CW from up.

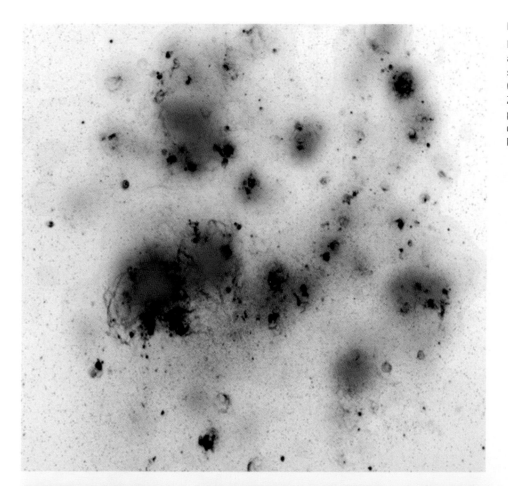

Figure 4.46

LMC: Optical Hα (greyscale, see also Figure 2.9) with <12.5 Myr star- formation (red) regions indicated. Credit: Original data from Harris and Zaritsky (2009). Reproduced by permission of the AAS. Hα image courtesy C. Smith, S. Points, the MCELS Team and NOAO/AURA/NSF.

Figure 4.47

LMC: Near-IR 1.2 μm (blue), 1.6 μm (green), 2.2 μm (red), 6.9° across.

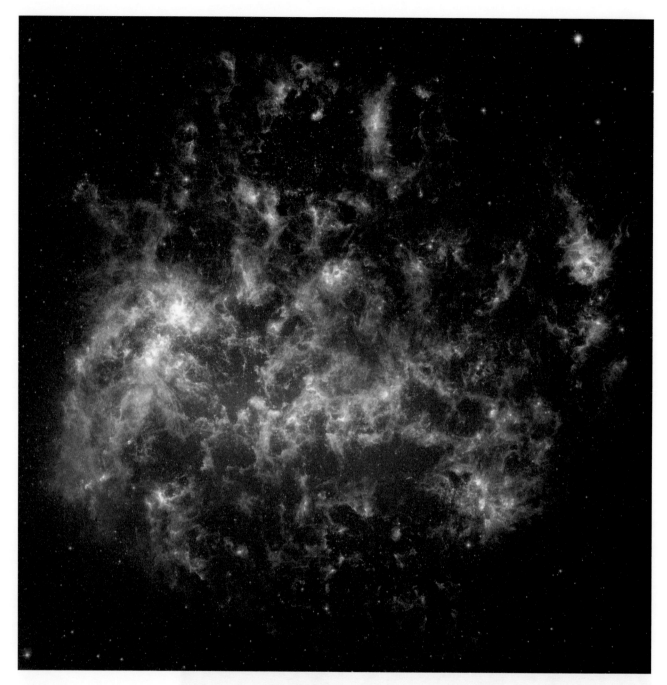

Figure 4.48

LMC: Near-IR 3.6 μm (blue) + mid-IR 8.0 μm (green), 24 μm (red), 7.4° across,
N is 27° CW from up.

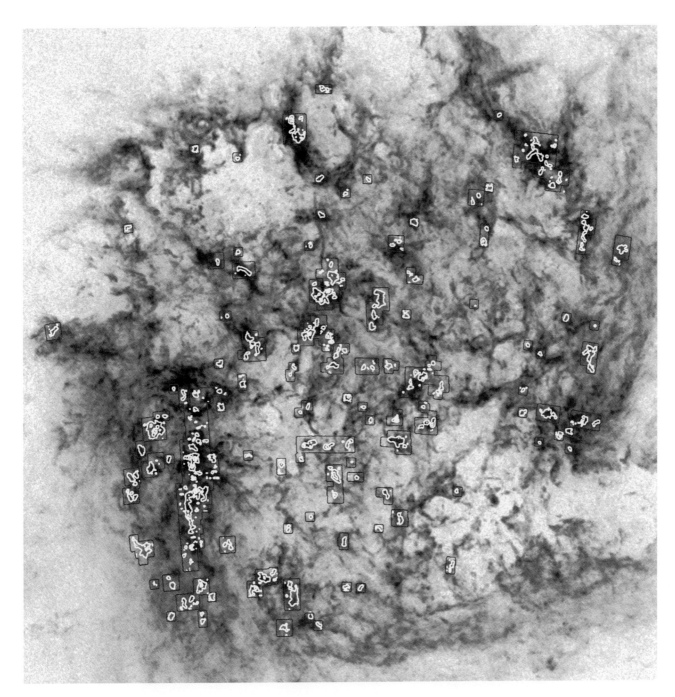

Figure 4.49

LMC: Radio CO ($J = 1 \rightarrow 0$) (white contours) overlayed on H I peak brightness
(a slightly smaller field of view than shown in Figure 1.25). Blue outlines indicate
the coverage of an ongoing CO survey; at present, the large survey area south of
30 Doradus (left, center) has received less integration time than other areas.
Credit: CO: A. Hughes (Swinburne), T. Wong (U. Illinois), J. Ott (NRAO), J. Pineda
(NASA/JPL), E. Muller (Nagoya. U.) and the MAGMA collaboration. H I:
L. Staveley-Smith.

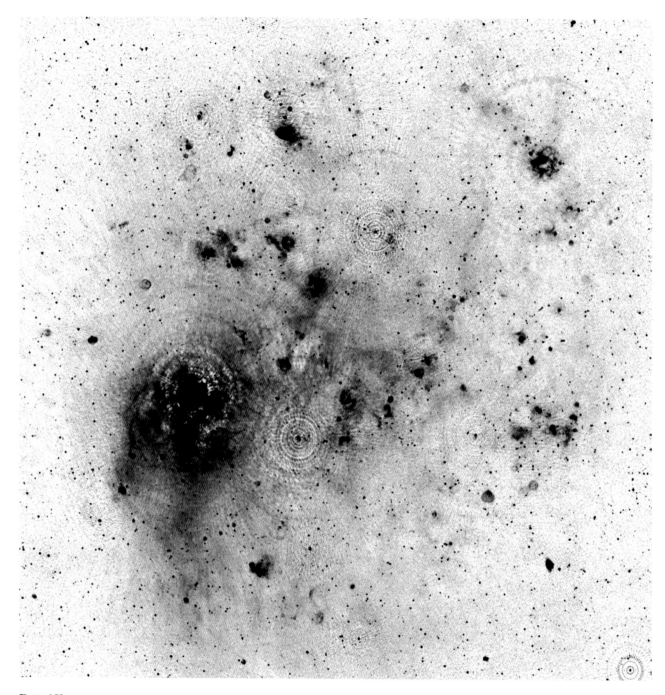

Figure 4.50

LMC: Radio 1.4 GHz continuum, rings are instrumental/calibration effects due to
bright sources. Credit: A. Hughes, L. Staveley-Smith and CSIRO.

4.1.8 NGC 2915

NGC 2915 is an isolated blue compact dwarf (BCD) galaxy, 3.3 Mpc distant.

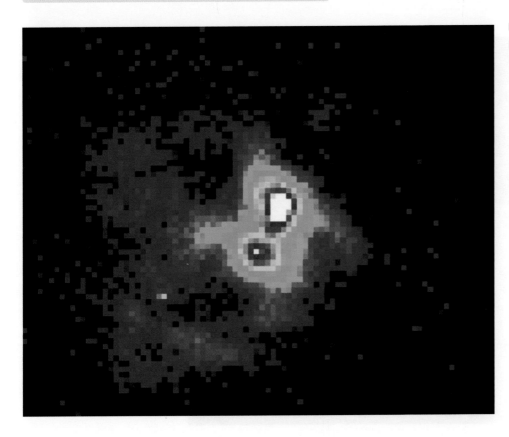

Figure 4.51
NGC 2915: Optical Hα, 3.6′ across.

■ NGC 2915:

The optical stellar population is described (Meurer, Mackie and Carignan 1994) by a concentrated high surface brightness blue population and a more extended red population.

Marlowe *et al.* (1995) present a CTIO 1.5 m telescope Hα image that shows bimodal central emission and two superbubbles with

diameters of ∼500 pc, with the north-east superbubble expanding at ±55 km s^{-1}.

Mass models derived from ATCA H I measurements (Meurer *et al.* 1996) indicate a total mass-to-(blue)light ratio (M_T/L_B) of 76, making this galaxy one of the darkest (largest dark matter) disk galaxies known.

Figure 4.52

NGC 2915: Optical R, 3.6′ across.

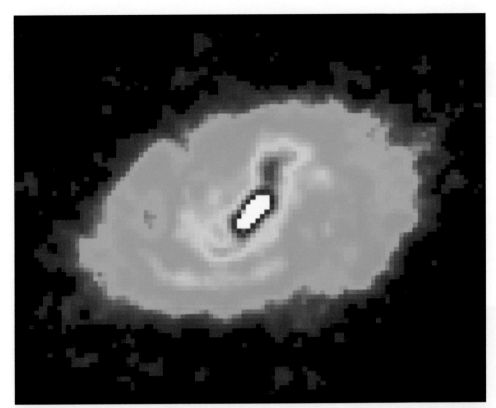

Figure 4.53
NGC 2915: Radio H I, 17 ′ across.

Figure 4.54
NGC 2915: Radio H I (blue), optical B (yellow), 17 ′ across.

4.1.9 Malin 2

Malin 2 (Bothun *et al.* 1990), also known as F568-6, is a low surface brightness (LSB) galaxy at a distance of 141 Mpc. These galaxies are defined as having central surface brightnesses fainter than the darkest night sky (~23 B mag. per sq. arcsec.). Whilst LSBs span the same range of physical parameters as brighter galaxies in the Hubble sequence, they are not low-mass dwarfs.

Figure 4.55

Malin 2: Optical Hα, 1$'$ across.

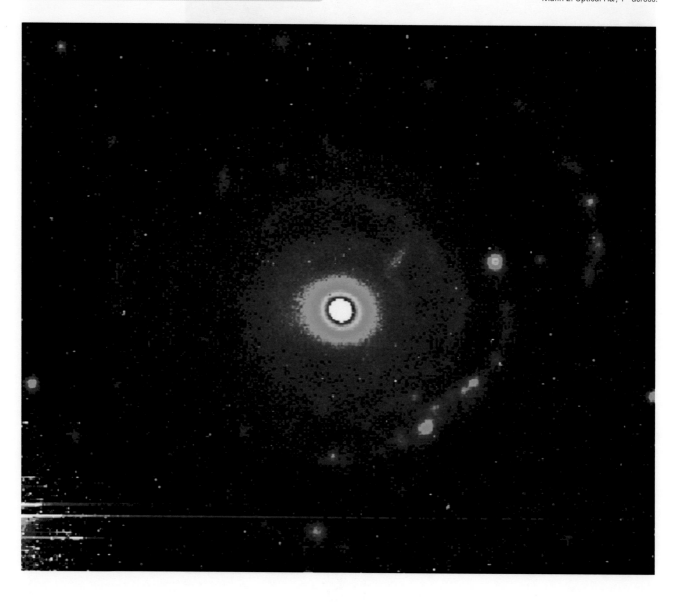

Malin 2:

McGaugh, Schombert and Bothun (1995) describe Malin 2 as a giant LSB disk. Spectroscopy of its H II regions shows metallicities higher than $0.3\odot$, suggesting an evolved stellar population.

Matthews, van Driel and Monnier-Ragaigne (2001) detect Malin 2 (F568-6) with 4×10^{10} M$_\odot$ of H I with the Nancay Radio Telescope.

Figure 4.56

Malin 2: Optical I.

4.1.10 NGC 5457/Messier 101

The famous "Pinwheel" Galaxy, NGC 5457 or Messier 101, is a Sc(s)I type, at a distance of 5.4 Mpc in the M 101 group. A grand design spiral, though asymmetric at large radii. The bright H II region NGC 5461 is in the southern spiral arm.

A secondary classification of I is adopted.

Figure 4.57

NGC 5457/M 101: X-ray softband 0.45–1.0 keV (purple), 1.0–2.0 keV (blue).

NGC 5457/M 101:

Snowden and Pietsch (1995) detect diffuse soft (0.1–0.3 keV) X-rays with ROSAT PSPC, indicative of thermal emission from 10^6 K gas. Pence *et al.* (2001) uses a deep CXO image to detect discrete X-ray sources and diffuse emission. The detection threshold is 10^{36} ergs s^{-1}, and the brightest source has a peak (super-Eddington luminosity[2]) of $>10^{39}$ ergs s^{-1}. Jenkins *et al.* (2005) use XMM-Newton to study the X-ray point source population. About 60% of sources are X-ray binaries (XRBs).

Bright H II regions are seen in the IRAS 60 μm image (Rice 1993).

Hippelein *et al.* (1996a) present ISO 60 and 100 μm images. Interestingly, at 60 μm the bright H II region NGC 5461 is brighter than the galaxy nucleus. An analysis of the 60 and 100 μm ISO data (Hippelein *et al.* 1996b) shows little color (dust temperature) change across the galaxy disk, and little correlation between far-IR and Hα fluxes, both in contrast to the results of Devereux and Scowen (1994) who used IRAS images.

A molecular gas bar was discovered by CO observations (Kenney, Scoville and Wilson 1991) that is offset ∼25° from the major axis of a slight oval distortion seen in the optical. The galaxy shows a distorted H I distribution at large radii, probably induced by tidal interactions with its neighboring galaxies, NGC 5474, NGC 5477 and Ho IV.

Kamphuis *et al.* (1991) use WSRT to elucidate the nature of a remarkable hole, or superbubble in the H I distribution. The superbubble has a diameter of 1.5 kpc and an expansion velocity of 50 km s^{-1}. The kinetic energy of the H I shell is several $\times 10^{53}$ erg, which requires approximately 1000 supernovae to energize it. Other holes are apparent in the H I emission, however the symmetric velocity structure and high expansion velocity of this superbubble is striking.

Smith *et al.* (2000) use VLA H I and UIT far-UV data to study the molecular gas content, finding that the ISM is dominated by molecular gas, out to 26 kpc. The decrease of H I at <10 kpc is due to an increase in the dust-to-gas ratio that allows H_2 to preferentially form more than atomic hydrogen.

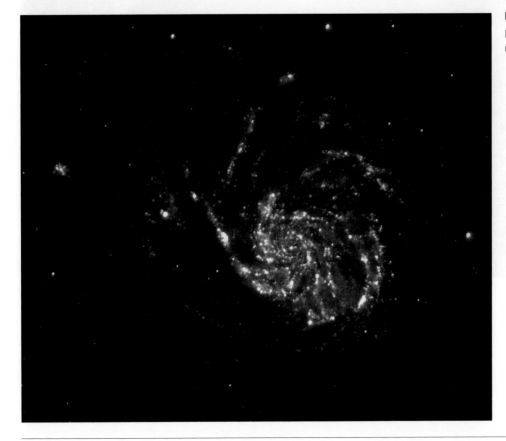

Figure 4.58

NGC 5457/M 101: Far-UV (blue), near-UV (red), 38′ across.

---

2 The Eddington luminosity or limit is the point where the gravitational force inwards equals the continuum radiation force outwards in a star, assuming hydrostatic equilibrium and spherical symmetry. To exceed this luminosity, and be stable, a star would need to have a stellar wind.

Figure 4.59

NGC 5457/M 101: Optical, 24 ′ across.

Figure 4.60

NGC 5457/M 101: Near-IR 3.6 μm (blue), mid-IR 8.0 μm (green), 24 μm (red),
24′ across, NGC 5461 is the bright region below and slightly left of the nucleus.
N is 30° CCW from up.

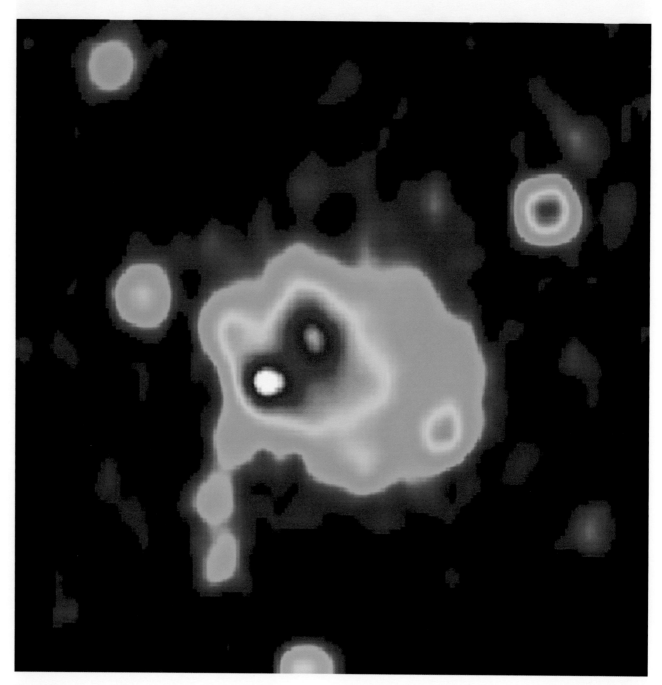

Figure 4.61

NGC 5457/M 101: Radio 6 cm continuum.

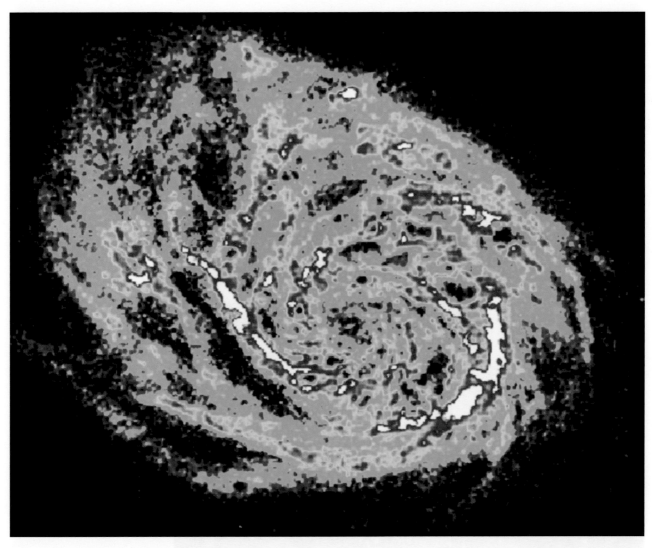

Figure 4.62
NGC 5457/M 101: Radio H I, 30′
across.

4.1.11 NGC 6822

NGC 6822 is an ImIV-V galaxy that was discovered by E.E. Barnard in 1884 and is also known as Barnard's Galaxy. It is a relatively isolated member of the Local Group, at a similar distance (0.7 Mpc) to NGC 224/M 31. It is asymmetrical in the sense that several dominant H II regions are seen at one end of its stellar distribution. It is obscured by the Galaxy due to its low galactic latitude of $b = -18.4°$.

A multicolor optical image of NGC 6822 is shown in Figure 1.19.

Figure 4.63

NGC 6822: Near-IR 1.2 μm (blue), 1.6 μm (green), 2.2 μm (red), 16′ across.

■ NGC 6822:

A study of the far-IR, H II regions and H I content by Gallagher *et al.* (1991) shows that ~50% of the far-IR flux has its origin near H II regions.

Wyder (2001) analyzes WFPC2 F555W and F814W photometry of the central bar to study the star-formation history. The best-fitting histories, based on color magnitude diagrams to faint magnitudes, suggest a constant or somewhat increasing star-formation rate from 15 Gyr ago to the present, except for a possible dip in the rate 3–5 Gyr ago.

Cannon *et al.* (2006) study the IR emission using Spitzer imaging. About 50% of the IR emission is associated with H II regions. The global dust-to-H I ratio is much lower than that derived for spiral galaxies.

IRAS 60 μm emission (Rice 1993) peaks are generally cospatial with H II regions.

The VLA 1.49 GHz continuum map (Condon 1987) shows several discrete sources that are cospatial with H II regions listed in Hodge (1977).

de Blok and Walter (2000) present H I imaging data that show an extended, elongated cold gas distribution with a large hole to the south-east of the optical galaxy. A tidal arm may be present, suggesting an interaction about 100 Myr ago.

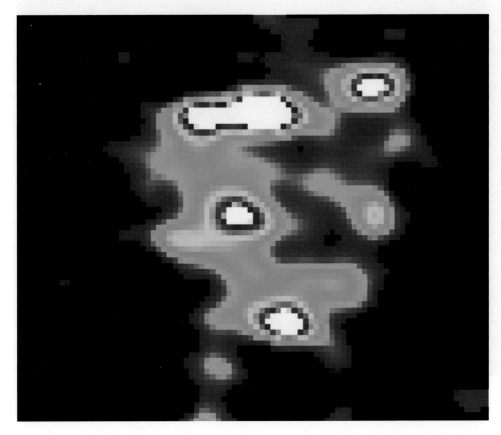

Figure 4.64

NGC 6822: Far-IR 60 μm, 30′ across.

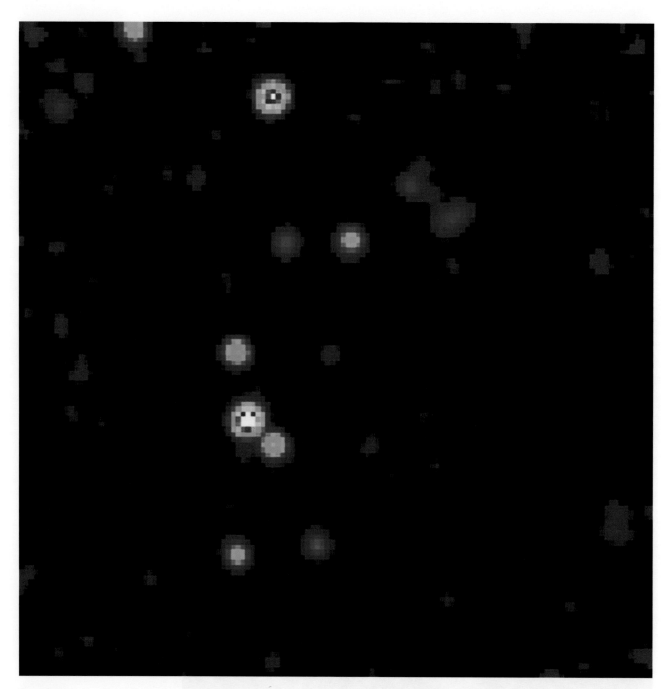

Figure 4.65

NGC 6822: Radio 1.49 GHz continuum, 30′ across.

4.2 INTERACTING GALAXIES

4.2.1 NGC 4406/Messier 86

NGC 4406 or Messier 86 is a bright Virgo Cluster $S0_1(3)$/E3 galaxy at 16.8 Mpc. It has a large velocity ($v_0 = -367$ km s^{-1}) relative to the Virgo Cluster ($v_0 = 1136$ km s^{-1} for the centrally located large elliptical NGC 4486/M 87), and is probably passing through the cluster core.

■ NGC 4406/M 86:

Einstein observations (Forman *et al.* 1979) discovered an X-ray plume extending to the north-west. Coupled with the unusual kinematics of the galaxy, this has promoted the suggestion that the hot ISM is being stripped by ram pressure[3] due to the intracluster medium (ICM). Nulsen and Carter (1987) detect excess optical emission near the X-ray plume and suggest it is a result of star formation due to cooling in the stripped hot ISM.

Trinchieri and di Serego Alighieri (1991) present a spectacular Hα + [N II] image from the European Southern Observatory 3.6 m telescope that shows filaments, arcs and rings predominantly on the SW side, that may be related to the unusual kinematical properties of the galaxy.

White *et al.* (1991) analyzed IRAS data and found a significant 60 μm component close to the position of the X-ray plume. The authors suggest that this emission is due to collisionally heated dust that has been exposed to the hot ISM of the galaxy due to the ram pressure stripping.

3 Ram pressure is caused by the drag force on a body travelling through a medium, in this case the ICM.

Figure 4.66
NGC 4406/M 86:
X-ray (blue),
optical (orange,
yellow). The
galaxy is at lower,
left. The image is
19′ vertically, N is
30° CCW from up.

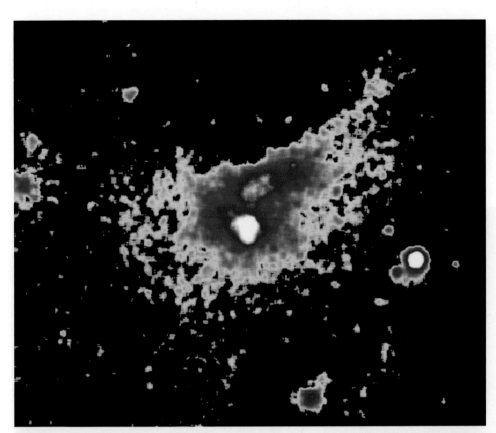

Figure 4.67

NGC 4406/M 86: Soft X-ray
(0.2–2.0 keV), 10′ across.

Figure 4.68

NGC 4406/M 86: Optical Hα + [N II],
200″ across.

4.2.2 NGC 4472/Messier 49

NGC 4472 or Messier 49 is an X-ray luminous Virgo Cluster E1/S0$_1$(1) galaxy located in a southern subclump of the cluster at 16.8 Mpc. It is the brightest optical galaxy in the Virgo Cluster, 0.30 mag. brighter than NGC 4486/M 87.

Figure 4.69

NGC 4472/M 49: X-ray (0.6–8.0 keV), 3.5′ across.

■ NGC 4472/M 49:

A ROSAT PSPC study (Forman *et al.* 1993) determined that the heavy element abundance of hot $\sim 10^7$ K gas was 1.5$\odot$ (well above solar) 4′–16′ from the nucleus. Biller *et al.* (2004) use ROSAT and CXO images and find X-ray holes or cavities that correspond with radio lobes. An X-ray tail, extending 8′ south-west, could exceed 100 kpc in extent.

McNamara *et al.* (1994) present optical and H I data providing strong evidence that an interaction between NGC 4472/M 49 and its dwarf neighbor, UGC 7636, has occurred. The H I and the stellar morphology of UGC 7636 appear similar.

Zepf *et al.* (2008) discuss the discovery of broad ($\sim$2000 km s^{-1}) [O III] lines in the globular cluster RZ 2109 of NGC 4472/M 49. To account for the emission, the [O III] is produced by photoionization across the cluster, suggestive of the presence of a stellar-mass black hole. This would be the first evidence of a black hole in a globular cluster.

A cloud of H I (Henning, Sancisi and McNamara 1993) is detected between NGC 4472/M 49 and UGC 7636. The H I has been removed from UGC 7636 by ram pressure due to the hot X-ray ISM of NGC 4472/M 49. The kinematics of the NGC 4472/M 49 nucleus shows the existence of a counter-rotating[4] stellar core.

Figure 4.70

NGC 4472/M 49: Near-IR 1.2 μm (blue), 1.6 μm (green), 2.2 μm (red), 5′ across.

---

4 A counter-rotating core rotates in the opposite direction to the majority of stars in the galaxy.

4.2.3 NGC 4676

NGC 4676 is commonly known as "The Mice" for obvious reasons. It consists of two galaxies, NGC 4676a (the northernmost galaxy) and NGC 4676b. It is 87 Mpc distant.

Figure 4.71

NGC 4676: Optical F475W (blue), F606W (green), F814W (red), 3.3′ across, N is ∼50° CW from up.

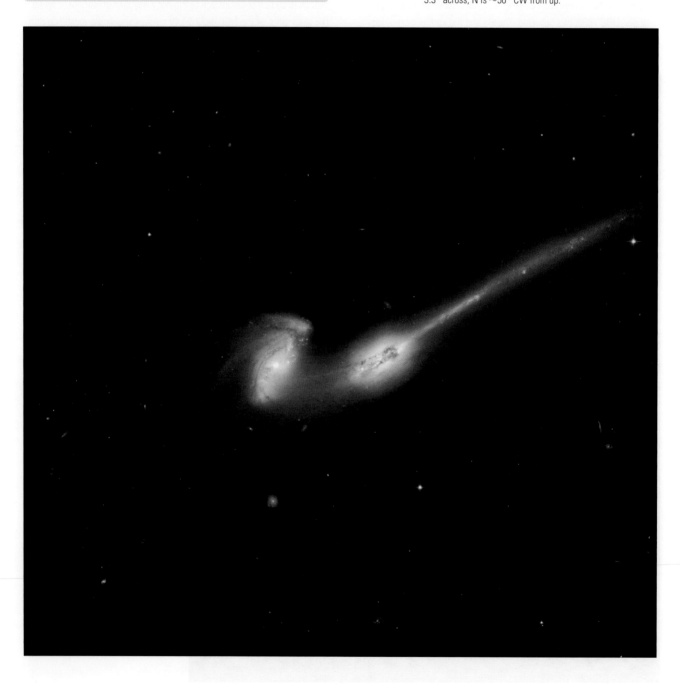

■ NGC 4676:

Originally modeled by Toomre and Toomre (1972) as a tidal (gravitational) encounter between two galaxies, this scenario is confirmed by subsequent kinematical observations. The tidal model has both galaxies in a prograde[5] collision, with their northern edges rotating away from us, and NGC 4676a at a larger redshift than NGC 4676b.

HST NICMOS J, H and K imaging is presented by Rossa *et al.* (2007).

The VLA H I tails are extensive, as is the H II emission, showing evidence for widespread star formation (Hibbard and van Gorkom 1996) induced by the encounter.

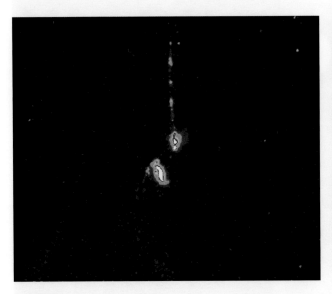

Figure 4.72
NGC 4676: Optical Hα + [N II], 6$'$ across.

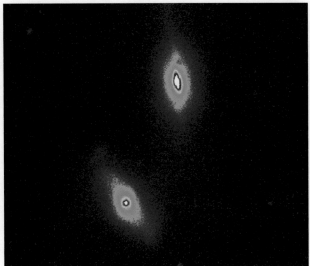

Figure 4.73
NGC 4676: Near-IR 2.2 μm.

---

5 The encounter is called prograde when the orbital spin angular momentum of the companion galaxy is aligned with the spin angular momentum of the target galaxy.

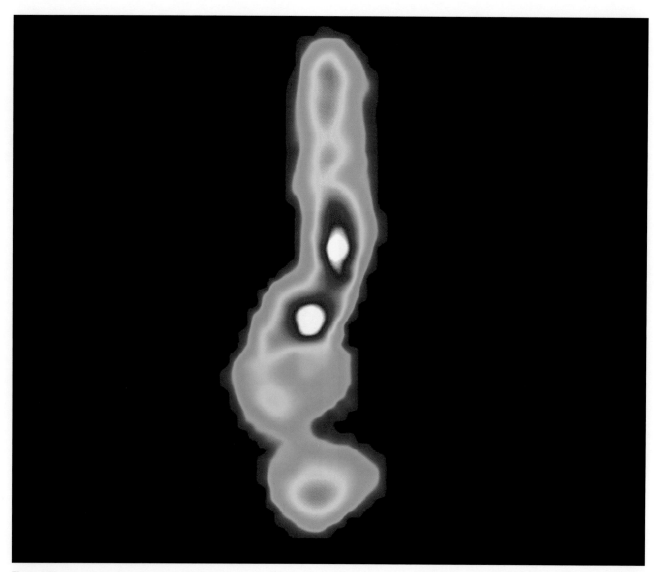

Figure 4.74

NGC 4676: Radio HI, 6′ across.

4.2.4 NGC 5194/Messier 51

NGC 5194 or Messier 51 is commonly known as the "Whirlpool" Galaxy. It is of type Sbc(s)I-II at 7.7 Mpc. This galaxy has been an important laboratory in the study of star-formation theory, especially testing the validity of density wave[6] theory that seeks to explain spiral arm structure in disk galaxies. To the north is its neighbor, NGC 5195, an $SB0_1$ pec galaxy that is interacting with NGC 5194/M 51 as suggested by the joining spiral arm.

An X-ray, UV, optical and IR image of NGC 5194/M 51 is shown in Figure 1.5.

An image of NGC 5194/M 51 at 850 μm (submillimeter region) is shown in Figure 2.11.

An optical image with the radio continuum (white contours) of NGC 5194/M 51 is shown in Figure 2.22. Magnetic field (yellow vectors) strength is also shown.

A secondary classification of A is adopted.

■ NGC 5194/M 51:

Ehle, Pietsch and Beck (1995) use ROSAT HRI and discover several X-ray point sources superimposed on diffuse emission that connects to NGC 5195. Terashima and Wilson (2001) present CXO ACIS-S data that show the nucleus, southern extranuclear cloud, and northern loop. The X-ray spectrum of the nucleus shows a strong Fe Kα emission line at 6.45 keV.

The HST FOC mid-UV image (Figure 4.77; Maoz *et al.* 1996) shows an X-shaped dust lane across the nucleus and several compact sources at larger radii.

VLA radio continuum observations of the nucleus (Figure 4.86) by Crane and van der Hulst (1992) show a nuclear jet-like feature suggesting the existence of a low-luminosity active nucleus. Broad wings in nuclear optical emission lines suggest the galaxy should be classified as a LINER–Seyfert transition object, although the nucleus does not harbor a luminous compact X-ray source.

VLA H I observations by Rots *et al.* (1990) show a confused velocity structure that is due to warps, density wave induced streaming motions along spiral arms, and tidal features. The inner H I traces the star-forming regions of the spiral arms, whilst an outer H I tail begins in the south then extends 90 kpc to the east, then north.

Rand, Kulkarni and Rice (1992) compared H I, CO (from Owens Valley Radio Observatory), Hα and B images to show that the warm and cold gas distributions support the picture in which H I is largely a product of dissociation[7] induced by radiation from star-forming regions.

---

6 A wave that passes through a medium and compresses material whilst not changing the average position of material significantly. In this way spiral arms are a result of the passage of a wave that induces star formation.

7 A molecule can be destroyed by a photon to form another molecule and/or component atoms.

Figure 4.76

NGC 5194/M 51: Soft X-ray (0.1–2.4 keV).

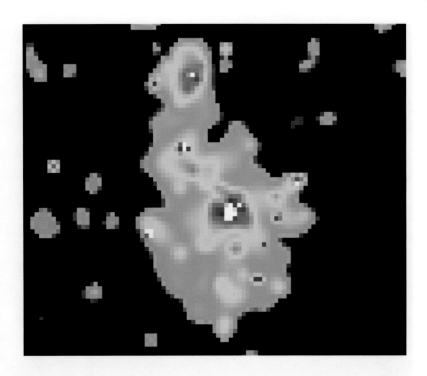

Figure 4.77

NGC 5194/M 51: Mid-UV 2800Å– nucleus, 10 ″ across.

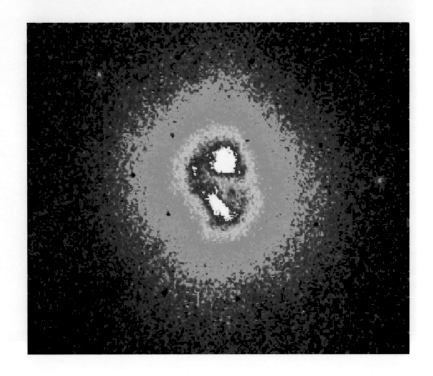

Figure 4.75   (opposite page)

NGC 5194/M 51: X-ray (0.1–10.0 keV), 7.7 ′ × 11.6 ′.

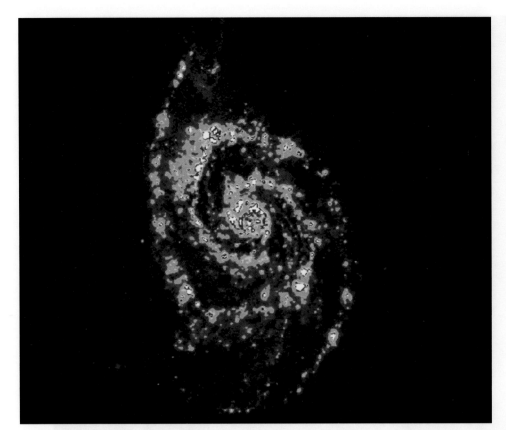

Figure 4.79
NGC 5194/M 51: Optical Hα.

Figure 4.80
NGC 5194/M 51: Optical – nucleus.

Figure 4.81
NGC 5194/M 51: Optical F439W (blue), F555W (green), (Hα + F814W red) – central.

Figure 4.82
NGC 5194/M 51: Near-IR 1.2 μm (blue), 1.6 μm (green), (2.2 μm+ Pα red) – central.

Figure 4.78 (opposite page)
NGC 5194/M 51: Optical, 9.6′ across.

Figure 4.84

NGC 5194/M 51: Radio CO 2.6 mm.

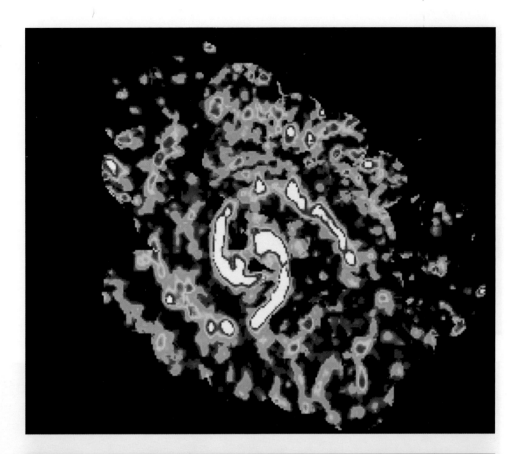

Figure 4.85

NGC 5194/M 51: Radio 4.86 GHz
continuum, 10′ across.

Figure 4.83   (opposite page)

NGC 5194/M 51: Near-IR 3.6 μm
(blue), 4.5 μm (blue-green), 5.6 μm
(yellow) + Mid-IR 8.0 μm (red), 9.9′×
13.7′.

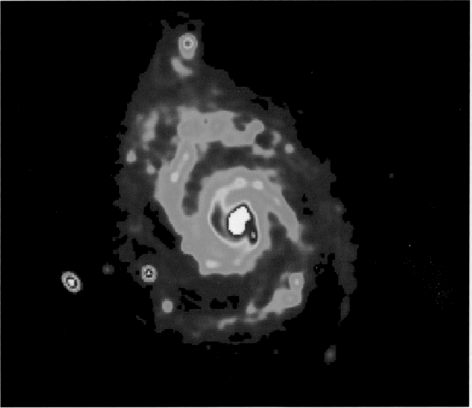

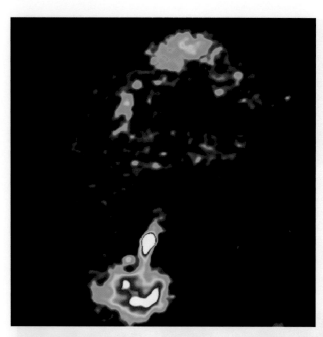

Figure 4.86

NGC 5194/M 51: Radio 4.86 GHz continuum – nucleus, 20 ″ across.

Figure 4.87

NGC 5194/M 51: Radio H I – intensity, 35 ′ across.

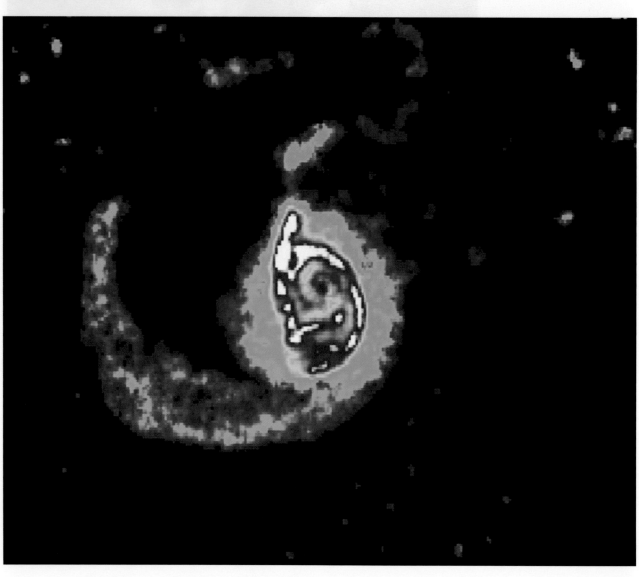

4.3 MERGING GALAXIES

4.3.1 NGC 520

NGC 520 is a morphologically disturbed amorphous galaxy some 27.8 Mpc away. Early studies suggested that its unusual morphology could be explained by a single galaxy undergoing an explosion. Recent observations, however, all favor a merger scenario.

An optical image is shown in Figure 1.20.

Figure 4.88

NGC 520: Optical Hα + [N II], 100″ across.

■ NGC 520:

CXO X-ray imaging by Read (2005) shows that the more massive, south-eastern nucleus dominates the emission – via a nuclear X-ray source and diffuse emission – that is not seen in the secondary nucleus. The total X-ray emission appears to be a factor of 2 less than expected for its evolutionary (merger) stage.

Near-IR imaging (Stanford and Balcells 1990; Bushouse and Stanford 1992) decreases the confusing effects of large amounts of dust, and detects two remnant mass concentrations. HST NICMOS J, H and K imaging is presented by Rossa *et al.* (2007).

VLA observations of Hibbard and van Gorkom (1996) show an outermost ring of H I extending through the nearby dwarf galaxy UGC 957, although it is unlikely (Stanford and Balcells 1991) that the main features of NGC 520 are caused by the passage of UGC 957.

Hibbard and van Gorkom (1996) surmise that NGC 520 probably formed from an encounter between a gas-rich disk galaxy and a gas-poor S0 or Sa. Hibbard, Vacca and Yun (2000) suggest that the anticorrelation seen between the spatial distributions of the gaseous and stellar tidal features may be explained by a starburst superwind.

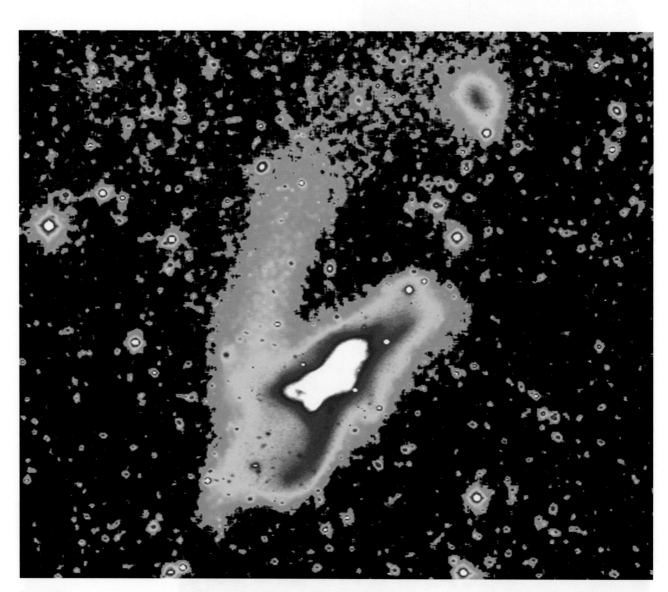

Figure 4.89

NGC 520: Optical R, interacting nearby galaxy UGC 957 is at top, right, 11 ′ across.

Figure 4.90
NGC 520: Near-IR 2.2 μm – central.

Figure 4.91
NGC 520: Radio H I, 12′ across.

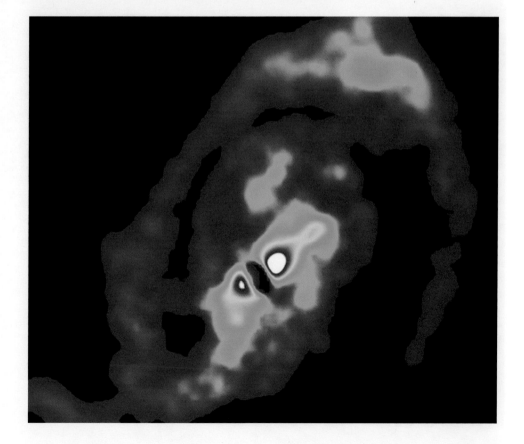

4.3.2 NGC 1275

NGC 1275, of E pec type, is the dominant galaxy in the Perseus (Abell 426) Cluster about 72.4 Mpc distant. This galaxy has numerous interesting properties that have prompted many observational studies. It possesses an active nucleus (Seyfert 1943) that is closer to Sy 2 classification than Sy 1. It has strong Hα emission and large quantities of atomic and molecular gas. It is a strong core-dominated radio source Perseus A (3C 84), and its hot, 10^7 K gaseous halo is cooling in the inner $\sim$100 kpc at a large rate of $\sim$200 $M_\odot$ yr^{-1}. Early studies showed two distinct ($\Delta v \sim$3000 km s^{-1}) emission line systems, with the stellar body associated with the lower velocity gas. Whilst observations cannot uniquely determine a single reason for its complex properties (McNamara, O'Connell and Sarazin 1996), it is obvious that several isolated or related phenomena are currently taking place.

A secondary classification of A is adopted.

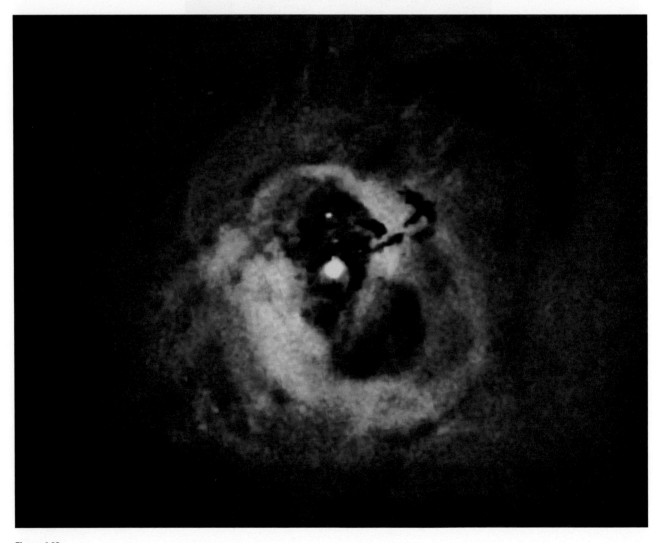

Figure 4.92

NGC 1275: X-ray (0.3–7.0 keV), 3.9′ across.

◼ NGC 1275:

Abdo *et al.* (2009) detect high-energy (>100 MeV) gamma-ray emission, possibly indicating year-to-decade time-scale variability.

The HST FOC mid-UV image (Maoz *et al.* 1995) shows a bright nuclear source that is unresolved (FWHM < 0.05"), which would suggest a maximum physical size of ~10 pc.

McNamara, O'Connell and Sarazin (1996) detect blue optical light indicative of recent star formation along the northern radio lobe that may be due to shock compression of gas by the radio emission. However, the inferred cooling rate of the hot X-ray gas is also similar to the recent rate of star formation over the last few 100 Myr.

HST imaging observations (Holtzman *et al.* 1992) that show the presence of blue, compact objects (young star clusters) near the nucleus also support a recent merger.

Figure 4.93

NGC 1275: Optical F435W (blue), F550M (green), F625W (red), 4′ across.

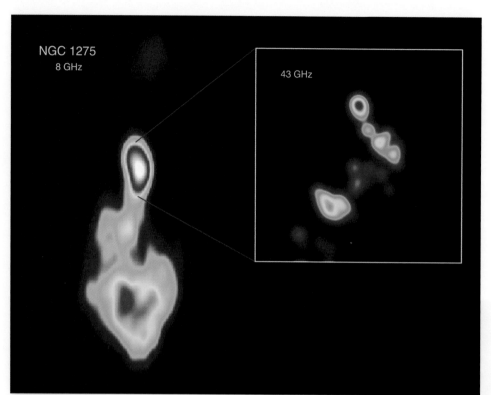

Figure 4.94

NGC 1275: Radio 43 and 8 GHz continuum, 30″ vertically (8 GHz).

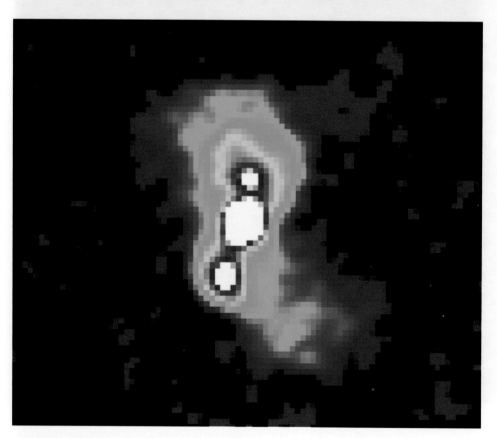

Figure 4.95

NGC 1275: Radio 90 cm continuum, 100″ vertically.

4.3.3 NGC 1316

NGC 1316 is a Sa pec (merger?) galaxy and is one of the best examples of a galaxy undergoing a merger event. It is located in the outskirts of the Fornax Cluster at 16.9 Mpc. The smaller barred spiral galaxy NGC 1317 is north of NGC 1316 with a $\sim$200 km s^{-1} larger velocity. It is not clear whether NGC 1317 is interacting with the larger NGC 1316; however UV observations of NGC 1317 show a circumnuclear ring centered on the barred nucleus. Such rings may be associated with the bar, or by a gravitational impulse possibly from NGC 1316.

The optical properties of NGC 1316 have been extensively investigated by Schweizer (1980, 1981). The galaxy possesses an $r^{1/4}$ (elliptical-like) surface brightness profile (see Section 4.3.5) that indicates dynamical[8] relaxation, yet at large radii, tidal tails, shells, ripples and loops (all signatures of ongoing interactions and merger events) are seen. It is an active galaxy of LINER type. Fornax A, the associated radio source, shows two lobes of emission either side of the optical galaxy.

An optical and radio image of NGC 1316 is shown in Figure 1.21.

A secondary classification of A is adopted.

■ NGC 1316:

ROSAT PSPC observations (Feigelson et al. 1995) detect X-ray inverse Compton emission from scattering of the cosmic microwave background by relativistic electrons in the radio lobes. ROSAT HRI observations (Kim, Fabbiano and Mackie 1998) show the inner $\sim$40″ X-ray emission is elongated along a direction that is perpendicular to the direction of the radio jets. Morphologies of gas emission suggest a different origin for the hot and cold ISM. Kim and Fabbiano (2003) use CXO imaging data to confirm the presence of cavities in the ISM related to the radio jets, a mixing of cold and hot ISM in the central regions consistent with a recent merger, and detail the point source population.

HST FOC imaging (Fabbiano, Fassnacht and Trinchieri 1994) detects a UV-bright, unresolved nucleus.

HST WFPC observations (Shaya et al. 1996) find that the star cluster population is quite normal, in contrast to that seen in another merger galaxy, NGC 1275 (Holtzman et al. 1992). Mackie and Fabbiano (1998) present deep CTIO Curtis Schmidt optical images and archival ROSAT PSPC images. An elliptical optical light model is subtracted, revealing extensive low surface brightness tidal tails, shells and loops. X-ray emission is cospatial with two tidal tails and X-ray spectra for one of the regions indicate soft emission, indicative of 5×10^6 K gas. The discovery of a large amount of warm, 10^4 K gas (denoted EELR in Figure 4.99) projected inside the larger tidal tail (L_1; Schweizer 1980) suggests it is related to a recent, $\sim$0.5 Gyr old, merger event. Grillmair et al. (1999) describe the luminosity and B − I color distribution of globular clusters using WFPC2 data. They suggest that these clusters are more like Galactic old open clusters than typical globular clusters in early-type galaxies.

Horellou et al. (2001) present CO and H I observations. CO is detected in the nuclear region, whilst H I was detected, associated with EELR (Mackie and Fabbiano 1998) and the giant H II region SH2 (Schweizer 1980).

The radio source, Fornax A, PKS 0320 − 374, associated with the galaxy has a double lobe structure separated by $\sim$33′ (Ekers et al. 1983) and exists outside the extent of the optical galaxy. A steep spectrum radio core with dual-opposing jets (Geldzahler and Fomalont 1984) exists within 1′ radius.

Polarization has been discussed in Section 2.8.5. The filamentary structure in the left (eastern) lobe is due to depolarization occurring in the lobe. The well-defined channel across the north of the western lobe shows depolarization caused when the position angle changes markedly.

---

8 Dynamical relaxation means that orbits are stable and ordered.

Figure 4.96

NGC 1316: Far-UV 1400–1700 Å (blue), near-UV 1800–2750 Å (green), optical (red),
15′ across.

Figure 4.97

NGC 1316: Optical B, NGC 1317 is N of NGC 1316, 22′ across.

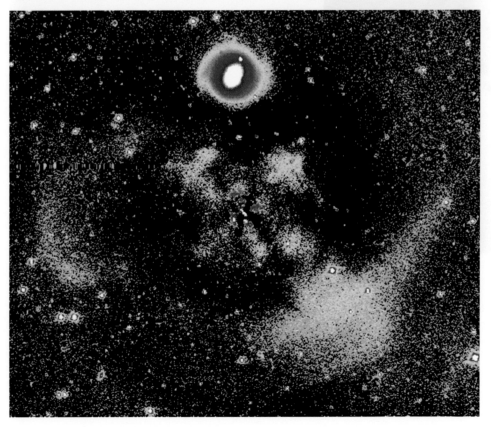

Figure 4.98

NGC 1316: Optical B, model subtracted. The bright "clover-leaf" red and white features near the nucleus are due to the model subtraction and are not real.

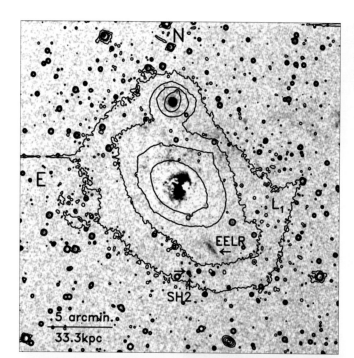

Figure 4.99

NGC 1316: Optical Hα + [N II] (greyscale), B contours of 23, 24, 25 and 26 mag arcsec^{-2} from Mackie and Fabbiano (1998). Extended emission-line region (EELR), previously known giant H II region (SH2; Schweizer 1980) and tidal tail (L$_1$; Schweizer 1980) are indicated. Reproduced by permission of the AAS.

Figure 4.100

NGC 1316: Optical – F435W (blue), F555W (green), F814W (red) – nucleus, 2.7′ across, N is ∼10° CW from up.

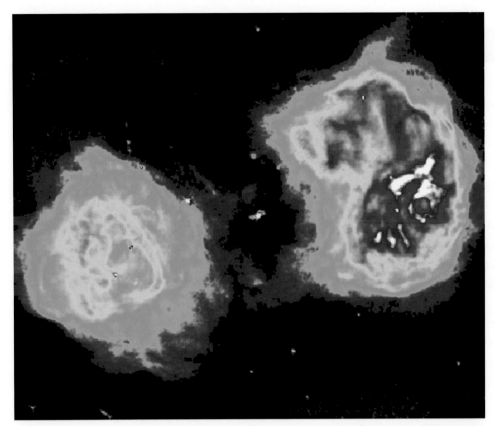

Figure 4.101
NGC 1316: Radio 1.51 GHz continuum – total intensity, 1° across.

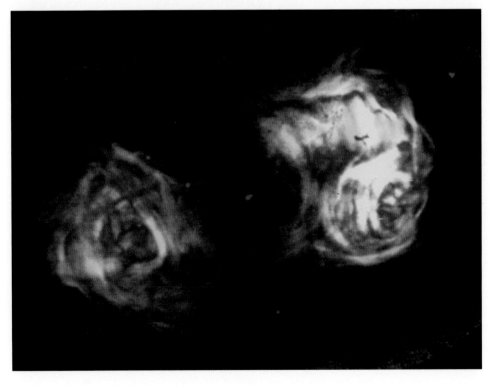

Figure 4.102
NGC 1316: Radio 1.4 GHz continuum – polarization. High polarization (white), high de-polarization (red).

4.3.4 NGC 4038/NGC 4039

NGC 4038/NGC 4039 are aptly named "The Antennae". They are of type Sc pec and are a classic example of two merging disk galaxies (NGC 4038, the northernmost galaxy and NGC 4039). Gravitational tidal forces have "drawn out" two long tails of stars and gas that resemble antennae. At 25.5 Mpc distance, the two disks are clearly separated and two tidal tails each extend over 100 kpc (or roughly two Galaxy diameters). Numerical simulations suggest that the two spirals started to merge ~700 Myr ago.

Figure 1.10 shows a large field of view optical image and Figure 1.11 a HST WFPC2 optical image of the two merging disks of NGC 4038/9.

Figure 2.5 shows an X-ray image of the central region of the two disks.

■ NGC 4038/9:

ROSAT PSPC observations (Read, Ponman and Wolstencroft 1995) show that 50% of the X-ray flux is associated with the disks and H II regions within, whilst the remainder of the emission comes from a halo of hot diffuse gas. Fabbiano, Schweizer and Mackie (1997) use ROSAT HRI with its greater angular resolution than PSPC to detect X-ray emission in the disks. X-ray emission is primarily cospatial with the H II regions, and shows complex filamentary structures suggesting both nuclear outflows and superbubbles. Fabbiano *et al.* (2003) use CXO to study the diffuse soft emission, which is complex, displaying different temperatures and perhaps non-solar abundances.

Hibbard *et al.* (2005) present GALEX UV imaging data – the UV generally follows the structure of the H I tidal tails. The UV suggests that most of the UV radiation comes from stars older than the tidal tails themselves, although some UV-bright regions do show more recent star formation.

HST NICMOS J, H and K imaging is presented by Rossa *et al.* (2007).

Vigroux *et al.* (1996) present ISO 6.7 and 15 μm imaging data, suggesting that emission longward of 12.5 μm is dominated by ionized gas and by reprocessing of UV flux by dust. The overlap region between the two disks is the most active star-formation region and dominates the total infrared luminosity between 12.5 and 18 μm.

CO observations by Stanford *et al.* (1990) show large amounts of molecular gas at the nuclei of each disk and at the contact region of the disks.

Hibbard *et al.* (2001) present VLA C+D array H I observations and confirm that the northern tidal tail has H I along its outer length but none at its base. They suggest massive stars in the disk of NGC 4038 may have ionized any such original gas. The H I in the southern tidal tail has a bifurcated structure, with one filament lying along the optical tail and another with no optical counterpart running parallel to it.

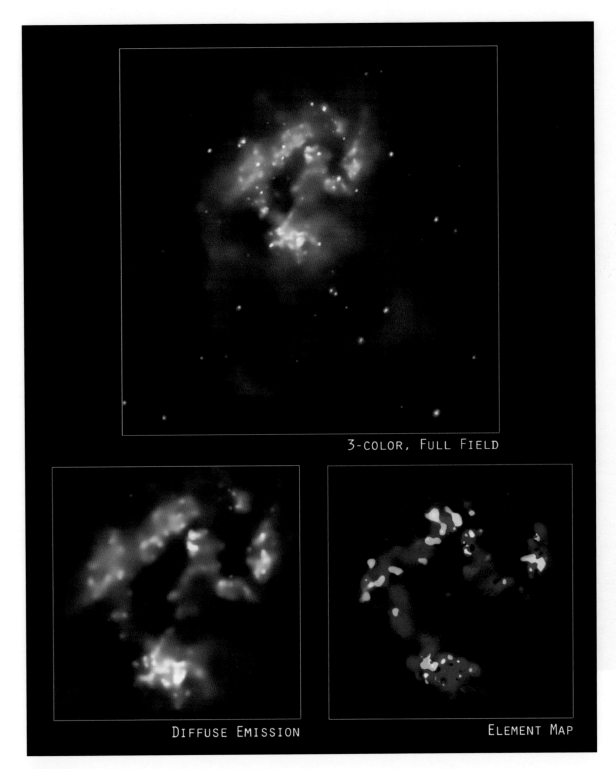

3-COLOR, FULL FIELD

DIFFUSE EMISSION

ELEMENT MAP

Figure 4.103

NGC 4038/9: X-ray:
top panel – 0.3–0.65 keV (red), 0.65–1.5 keV (green), 1.5–6.0 keV (blue), 4.8′
across; bottom, left – diffuse emission; bottom, right – element map: iron (red),
magnesium (green), silicon (blue), 2′across.

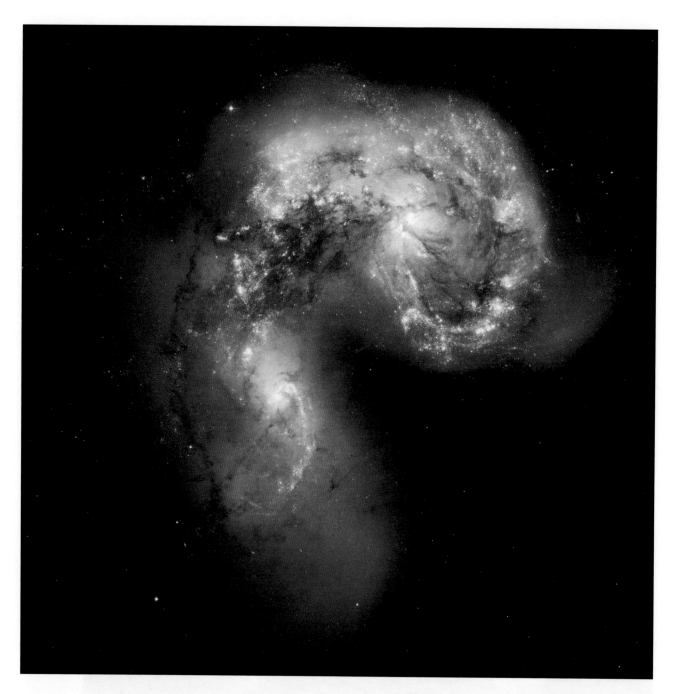

Figure 4.104

NGC 4038/9: Optical F435W (blue), F550M (green), F658N (Hα+[N II]) (pink), F814W (red), 3.2′ across, N is ~20° CW from up.

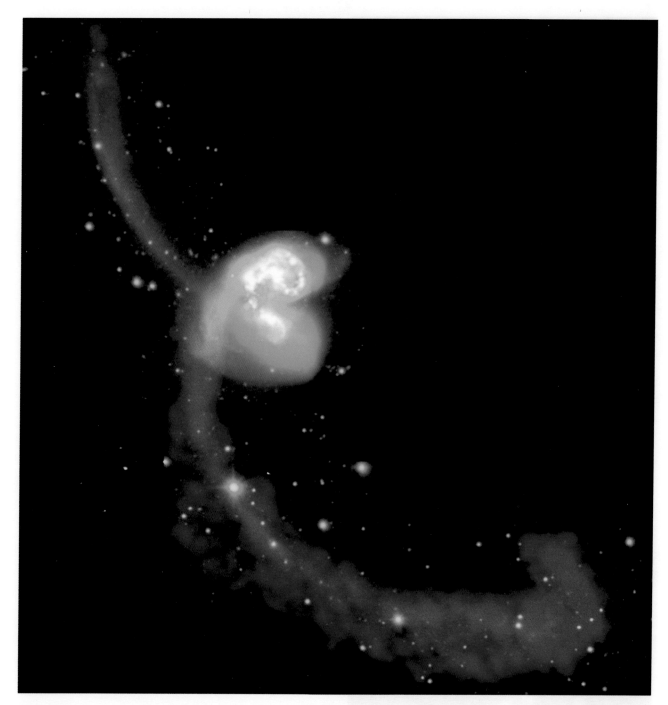

Figure 4.105

NGC 4038/9: Optical (green) + radio
H I (blue).

Figure 4.106

NGC 4038/9: Near-IR 1.2 μm (blue), 1.6 μm (green), 2.2 μm (red), 4′ across.

Figure 4.107

NGC 4038/9: Near-IR 3.6 μm (blue), 4.5 μm (green) + near/mid-IR 5.8 + 8.0 μm (red), 4.3′ across, N is 5.5° CW from up.

4.3.5 NGC 7252

NGC 7252 is a merger galaxy, sometimes called "Atoms for Peace"[9] due to its distinctive internal structure of loops and shells. It is at a distance of 63.5 Mpc and is the prototypical example of an evolved merger remnant of two disk galaxies, displaying a single nucleus and tidal tails.

Figure 4.108

NGC 7252: Optical V+I, 30″ across.

9 This was the title of a speech by U.S. President Dwight Eisenhower to the UN General Assembly on December 8, 1953. The Atoms for Peace Award was established in 1955 to recognize the development or application of peaceful nuclear technology. In July, 1955, the U.S. Post Office issued an "Atoms for Peace" stamp that had two hemispheres of Earth surrounded by three electron-like "orbitals". The similarity of NGC 7252 with its loops and shells to this has prompted this name.

■ NGC 7252:

Extensive optical observations by Schweizer (1982) showed that NGC 7252 has a single nucleus with an elliptical-like light distribution. Analytic surface brightness profiles of the form

$$I \propto \exp(-r^{\alpha})$$

with I, the intensity, as a function of radius, r, tend to describe elliptical galaxy radial profiles reasonably well with $\alpha = 0.25$. This is commonly known as the $r^{1/4}$ law or de Vaucouleur's law after Gerard de Vaucouleurs who promoted its use. This version can be rewritten as

$$I(r) = I_{\mathrm{e}} \exp\left\{ -7.67 \left[\left(\frac{r}{r_{\mathrm{e}}}\right)^{1/4} - 1 \right] \right\}$$

where $I(r)$ is the surface brightness at radius r, and r_{e} is the radius in which half the light is enclosed. NGC 7252 has an $r^{1/4}$-like surface brightness profile, as predicted for evolved disk–disk mergers.

Two tidal tails extend 80 and 130 kpc from the center. The nucleus contains a small disk of ionized gas, and displays an A star type spectrum suggestive of recent star formation.

HST PC observations (Figure 4.108, Whitmore *et al.* 1993) detect point-like objects that are probably young globular clusters formed within the last 1 Gyr, the approximate time-scale of the merger event. A clear spiral pattern is observed around the nucleus that is probably short-lived.

The multiwavelength study by Hibbard *et al.* (1994) presents VLA H I and X-ray observations. Interestingly, all of the H I is found in the outer regions of the two extensive tidal tails. In contrast, the ROSAT PSPC X-ray emission is centered on the stellar spheroid. In addition to the presence of central warm, 10^4 K gas, and molecular gas in the nucleus (Dupraz *et al.* 1990), this suggests an efficient mechanism of H I conversion into other centrally located gaseous phases.

Figure 4.109

NGC 7252: Optical R, 10′ across.

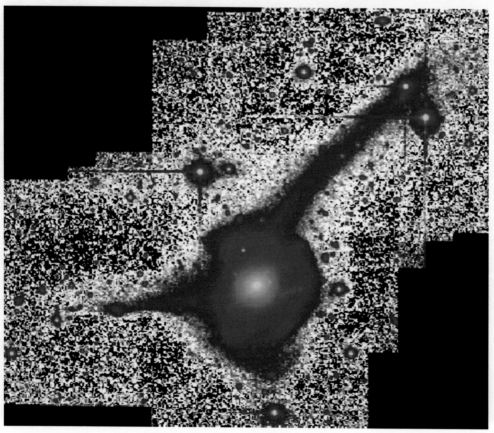

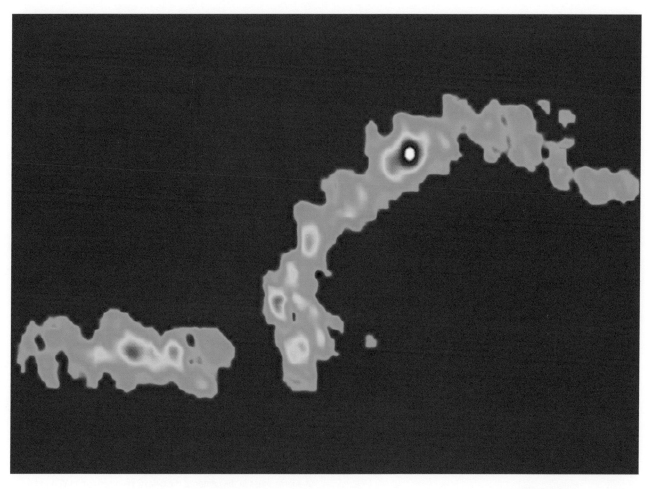

Figure 4.110
NGC 7252: Radio H I, 12′ across.

4.4 STARBURST GALAXIES

4.4.1 NGC 253

NGC 253 is a spectacular, nearly edge-on (inclination of approximately 70° to our line of sight), starburst spiral. It is 3 Mpc distant in the South Polar Group or Sculptor Group. NGC 253 is rapidly forming stars, at rates of one to two orders of magnitude greater than more normal spirals. It is ~20% as massive as the Galaxy or NGC 224/M 31.

■ NGC 253:

Itoh *et al.* (2002) detected TeV gamma-ray emission greater than 0.5 TeV with the CANGAROO-II 10 m imaging atmospheric Cerenkov telescope.

Pietsch (1994) and Vogler and Pietsch (1999) show ROSAT PSPC images. The (0.1–2.4 keV) X-ray image shows discrete sources (probably low- and high-mass X-ray binary systems), with a high concentration near the nucleus, as well as faint diffuse emission. The soft (0.1–0.4 keV) image shows spectacular diffuse emission extending above the plane of the galaxy. This emission is 10^{5-6} K gas being energized and ejected from the plane by SN explosions. Weaver *et al.* (2002) present CXO ACIS data and detect a heavily absorbed source ($>10^{39}$ ergs s^{-1}) of hard X-rays embedded within the nuclear starburst region.

OVRO CO observations (Canzian, Mundy and Scoville 1988) detect a large amount of molecular gas in the form of a central bar. The mass is $\sim 5 \times 10^8$ M$_\odot$ and displays rigid body rotation suggesting the existence of a large mass in the center of the galaxy. High-resolution VLA radio continuum images (Ulvestad and Antonucci 1997) show numerous compact sources in the inner 200 pc. Of the brightest sources, about half are dominated by thermal emission from H II regions. Koribalski, Whiteoak and Houghton (1995) use ATCA to study the H I distribution. An asymmetric distribution is seen in the outer regions, as well as a bar in the disk, a rotating ring and outflows.

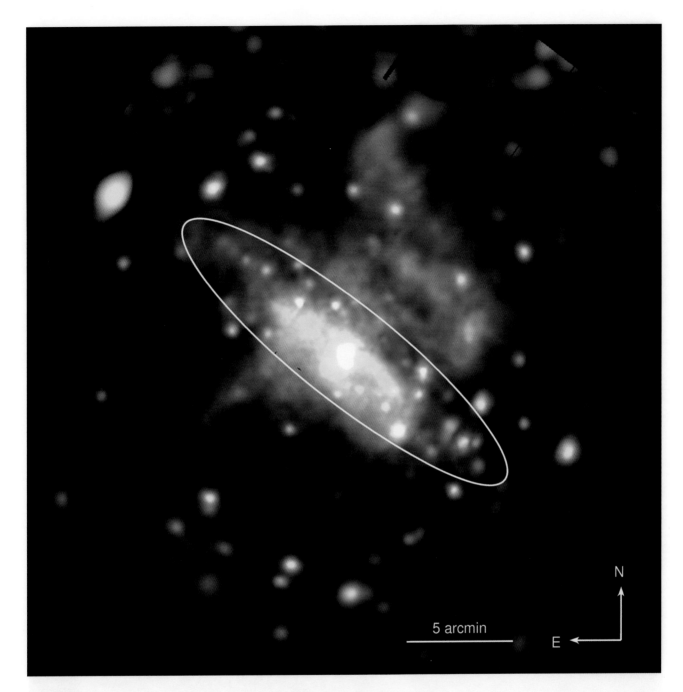

5 arcmin

N

E

Figure 4.111

NGC 253: Soft X-ray softband, 0.2–0.5 keV (red), 0.5–1.0 keV (green), 1.0–2.0 keV
(blue), 45′ across. The white line indicates the optical extent.

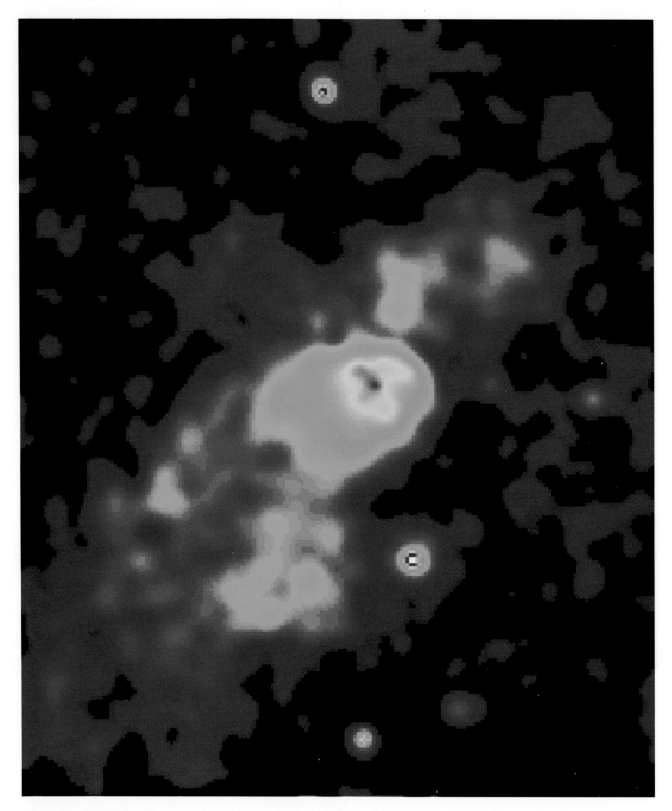

Figure 4.112

NGC 253: X-ray softband (0.2–1.5 keV), 2′ across.

Figure 4.113

NGC 253: Far-UV (1400–1700 Å) (blue),
near-UV (1800–2750 Å) (red),
22′ across.

Figure 4.114

NGC 253: Optical B – central.

Figure 4.115

NGC 253: Optical, F450W (blue), F555W (green), F814W (red), 250″ across,
N is ~140° CW from up.

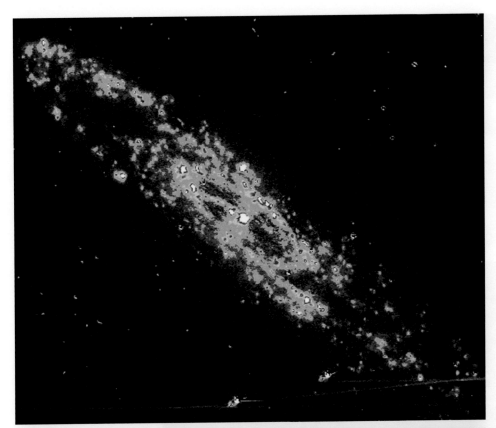

Figure 4.116
NGC 253: Optical Hα + [N II]
(CCD readout defects).

Figure 4.117
NGC 253: Near-IR 1.2 μm (blue),
1.6 μm (green), 2.2 μm (red), 19′
across.

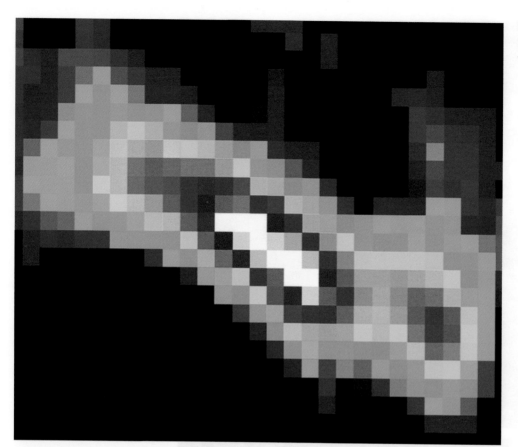

Figure 4.118

NGC 253: Radio CO 2.6 mm, 30 ″
vertically.

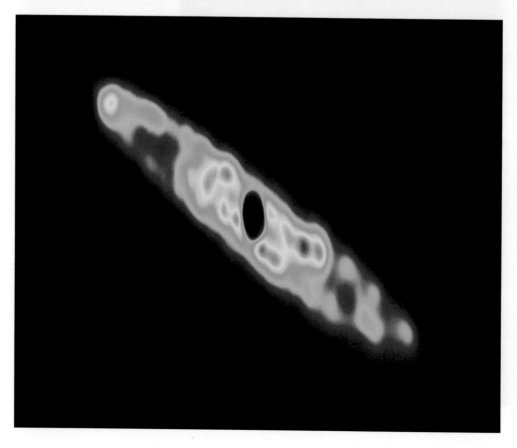

Figure 4.119

NGC 253: Radio H I, 20 ′
vertically. The image is copyright
CSIRO Australia 1995.

4.4.2 NGC 3034/Messier 82

NGC 3034 or Messier 82, 5.2 Mpc away, is classified as amorphous, and is the best studied starburst galaxy. It is located in the M 81 Group. Intense star formation is occurring in the central region which is also the site of warm ionized gas, far-IR and molecular gas emission. The galaxy was not highly luminous before the star-formation burst, making the present-day galaxy a unique laboratory since the starburst almost dominates the host galaxy. The nucleus appears singular at 2 μm but two peaks either side of the 2 μm peak are seen in Brγ, Hα and in the near- and mid-IR continuum.

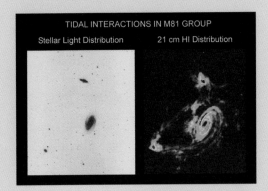

Figure 4.120

M 81 Group. Left: Optical image from left, NGC 3077 (bottom), NGC 3034/M 82 (top) and NGC 3031/M 81 (middle). Credit: Digital Sky Survey/STScI. Right: VLA H I map (Yun, Ho and Lo 1994). Credit: National Radio Astronomy Observatory/AUI/NSF.

H I observations shown in Figure 4.120 (Yun, Ho and Lo 1994) of NGC 3031/M 81 and NGC 3034/M 82 (the dominant members of the M 81 Group) and NGC 3077 show dramatic evidence of tidal interactions (filamentary features) between group members.

Radio continuum emission of NGC 3034/M 82 at 92.0 GHz is shown in Figure 2.12.

The radio/submillimeter/far-IR spectrum of NGC 3034/M 82 is shown in Figure 2.13.

A secondary classification of I is adopted.

■ NGC 3034/M 82:

ROSAT HRI observations (Bregman, Schulman and Tomisaka 1995) of the central region show several discrete X-ray sources in the nucleus, and much diffuse emission. The emission along the minor axis of the galaxy is consistent with gas outflows. Kong *et al.* (2007) discovered an ultraluminous X-ray (ULX) source in the CXO data. Its luminosity is as high as 1.3×10^{40} erg s^{-1} and appears to be located in a H II region. The authors suggest the ULX to be a >100 M$_\odot$ intermediate-mass black hole (IMBH).

Near-IR imaging with the Steward 2.3 m telescope presented by McLeod *et al.* (1993) has been used to constrain the properties of the starburst.

Yao (2009) describes starburst models using far-IR, sub-mm and mm line emission of molecular and atomic gas, irradiated by star clusters. Recent starburst activity is dated at 5–6 and 10 Myr.

Tilanus *et al.* (1991) present JCMT/BSISR CO ($J = 3 \to 2$) observations that show a double-peaked structure in the central regions, supporting the existence of a circumnuclear ring. The inferred ratio of ^{12}CO/^{13}CO indicates a similar abundance ratio to that observed in the Galaxy. Matsushita *et al.* (2000) compare CO and HCN interferometric observations of the central region with the Nobeyama Millimeter Array and have detected a molecular superbubble. The center of the superbubble is within the close vicinity of a purported massive >460 M$_\odot$ black hole (a second IMBH and different from that discussed by Kong *et al.* 2007) and the 2.2 μm secondary peak (a luminous supergiant-dominated cluster), which strongly suggests that these objects may be related.

Seaquist *et al.* (1996) compare radio recombination observations at H41α (92.0344 GHz) with 92 GHz continuum and HCO$^+$ (1–0) emission at 89.2 GHz. The H41α distribution appears clumpy compared to other ionized gas indicators; however, it shows a similarity to the HCO$^+$ (1–0) emission and thus the distribution of dense molecular gas. Seaquist, Lee and Moriarty-Schieven (2006) present a ^{12}CO ($J = 6 \to 5$) emission (691.5 GHz) map. They find the ($J = 6 \to 5$) emission in the starburst region more concentrated than the ($J = 1 \to 0$) emission, and excitation peaks in two spurs extending northward towards the superwind outflow.

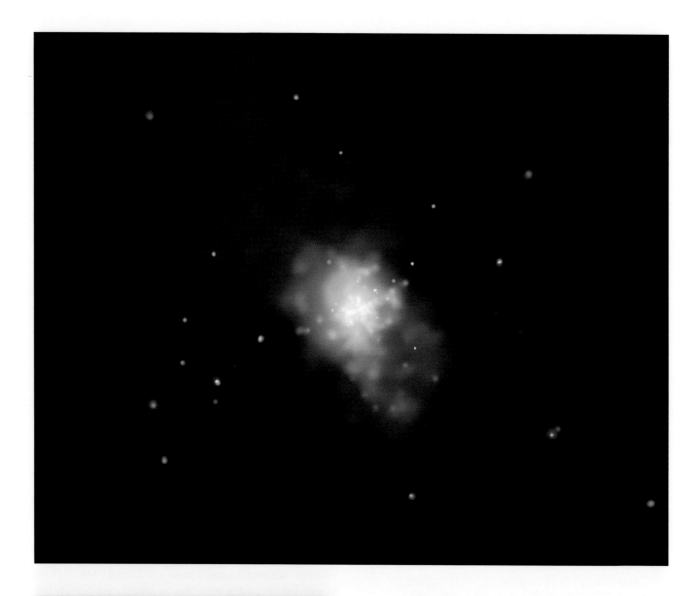

Figure 4.121

NGC 3034/M 82: X-ray (0.1–10.0 keV), low energy (red), intermediate (green), and high energy (blue), 7.9′ across, N is 40° CCW from up.

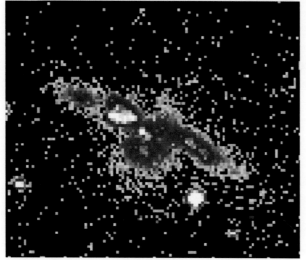

Figure 4.122

NGC 3034/M 82: Mid-UV 2800 Å, 8′ across.

Figure 4.123

NGC 3034/M 82: Optical B (blue), V (green), Hα (red), 7′ across. ©Subaru Telescope, National Astronomical Observatory of Japan (NAOJ) and is reproduced with permission.

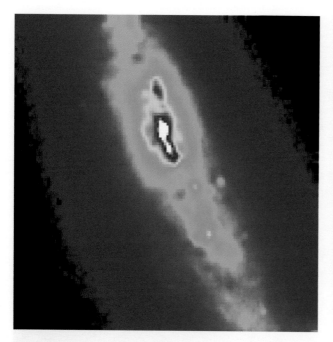

Figure 4.124

NGC 3034/M 82: Near-IR 1.6 μm, N is ~40° CW from up.

Figure 4.125

NGC 3034/M 82: Near-IR 3.6 μm (blue), 4.5 μm (green) + near/mid-IR 5.8 + 8.0 μm (red), 12.6′ across, N is 58° CCW from up.

Figure 4.126

NGC 3034/M 82: Radio CO, 1′ across.

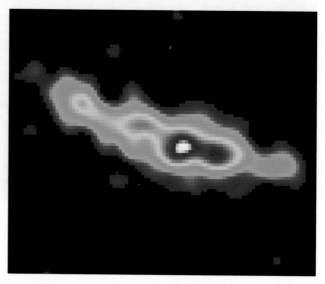

Figure 4.127

NGC 3034/M 82: Radio 0.3 cm continuum, 50″ vertically.

4.4.3 NGC 5236/Messier 83

An SBc(s)II galaxy, Messier 83, at an inclination of 24°, that displays a grand design two-armed structure. At a distance of 4.7 Mpc it is one of the finest spirals in the sky and is sometimes referred to as the "Southern Pinwheel". Its starburst nature may have been triggered by an earlier interaction with its companion galaxy, NGC 5253.

A secondary classification of I? is adopted.

■ NGC 5236/M 83:

Ehle *et al.* (1998) use ROSAT PSPC to detect 13 X-ray discrete sources, of which eight are cospatial with H II regions, suggesting an association of the X-ray sources with Population I sources. Diffuse emission has been modeled by two components, one associated with the halo, at 2×10^6 K, and a hotter disk component, at 6×10^6 K. Soria and Wu (2002) use CXO ACIS data and detect 81 point sources. A point source with $L_X \sim 3 \times 10^{38}$ erg s^{-1} in the 0.3–8.0 keV band coincides with the IR nuclear photometric peak. About 50% of the total emission in the nuclear region is unresolved. Strong emission lines are seen in the spectrum and the high abundances of C, Ne, Mg, Si and S with respect to Fe suggest that the interstellar medium in the nucleus is enriched and energized by type-II SN events and massive star winds.

Dong *et al.* (2008) present Spitzer imaging at 3.6, 4.5, 5.8 and 8.0 μm for two fields 19.5 kpc from the center of the galaxy. Combined with GALEX and H I data, star clusters in the outskirts of the galaxy are age-dated between 1 Myr and 1 Gyr, with a mean age of 180 Myr. The low-density stars (and gas) still seem to have formed following the laws favored in more high-density regions.

Muraoka *et al.* (2009) present CO ($J = 3 \rightarrow 2$) emission maps using the Atacama Submillimeter Telescope Experiment. An effective resolution of 25$''$ over an $8' \times 8'$ region allows detection of 54 giant molecular-cloud associations (GMAs) in arm and inter-arm areas. The inter-arm GMAs could be affected by shear motion, in which unvirialized GMAs with large velocity dispersions have star formation suppressed.

Tilanus and Allen (1993) compare VLA H I and optical Hβ observations and do not detect large (60–90 km s^{-1}) streaming velocities, suggesting that the galaxy possesses a small density wave. Tilanus and Allen (1993) also suggest that the H I ring and large extent of H I may be caused by a previous interaction.

Figure 4.128

NGC 5236/M 83: X-ray, 0.2–0.8 keV (red), 0.8–1.5 keV
(green), 1.5–12.0 keV (blue), 15′ across.

Figure 4.129

NGC 5236/M 83: Far-UV (1400–1700 Å) (blue), near-UV
(1800–2750 Å) (green), radio H I (red), ∼1° across.

Figure 4.130

NGC 5236/M 83: Optical, F336W (blue), F555W (green),
F814W (red), F502N ([O III]) (cyan), F657N (Hα)
(red) – nucleus is middle, right. The image is 2.6′ across,
N is ∼10° CCW from up.

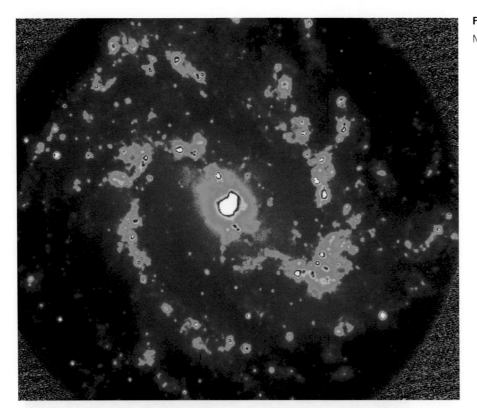

Figure 4.131
NGC 5236/M 83: Optical Hα.

Figure 4.132
NGC 5236/M 83: Optical I, 13′ across.

Figure 4.133

NGC 5236/M 83: Mid-IR 7 μm, N is
~50° CCW from up.

Figure 4.134

NGC 5236/M 83: Radio 6 cm continuum.

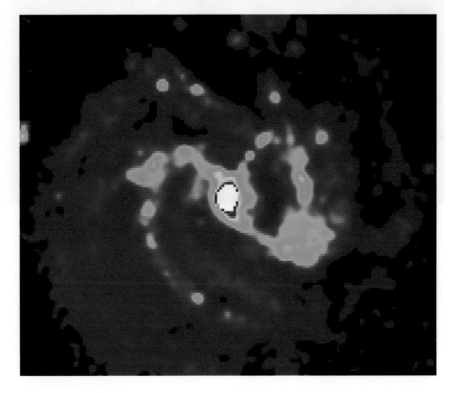

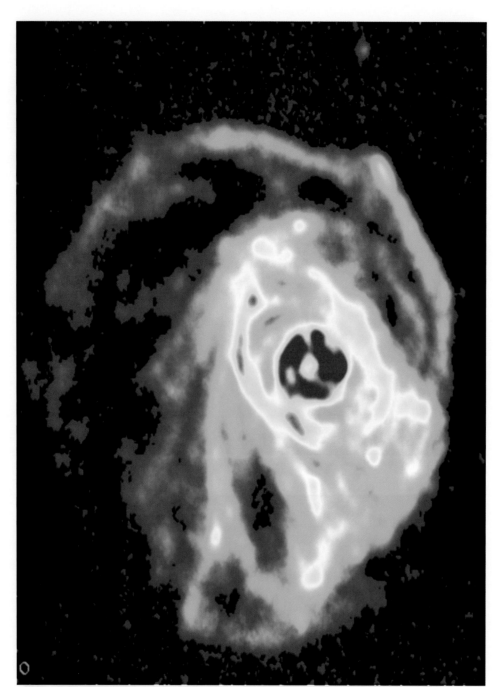

Figure 4.135
NGC 5236/M 83: Radio H I, 1.2°
vertically. The image is ©CSIRO
Australia 2004.

4.5 ACTIVE GALAXIES

4.5.1 NGC 1068/Messier 77

NGC 1068 or Messier 77 is the archetypal Seyfert 2 (Seyfert 1943) galaxy. It is located at 14.4 Mpc and is an Sb(rs)II type at an inclination of ∼30°.

Optical [O III] emission near the nucleus is shown in Figure 2.10.

■ NGC 1068/M 77:

ROSAT HRI imaging (Wilson *et al.* 1992) detects an unresolved nuclear X-ray source, and extended emission associated with the nucleus and the starburst disk. Young, Wilson and Shopbell (2001) present subarcsecond-resolution CXO imaging spectroscopy that shows a bright, compact source extended by 165 pc in the same direction as the nuclear optical line and radio continuum emission. Ogle *et al.* (2003) test the Seyfert unification theory using CXO data. The X-ray derived parameters of NLR column density, outflow velocity and electron temperatures are consistent with an obscured Seyfert 1 nucleus.

Analysis of polarized light by Antonucci and Miller (1985) shows an optical spectrum closely resembling a Seyfert 1 galaxy (see Section 1.5.5).

Narrowband emission line ([O III] 5007 Å) optical imaging shows high excitation gas distributed in a cone-like manner (Figure 2.10) to the NE of the nucleus (Evans *et al.* 1991). A stellar bar is seen in the near-IR.

OVRO CO ($J = 1 \rightarrow 0$) observations by Planesas, Scoville and Myers (1991) detect inner gaseous spiral arms originating from the ends of the stellar bar, and a compact nuclear source cospatial with the Seyfert 2 nucleus.

Wilson and Ulvestad (1983) present 6 cm/4.9 GHz VLA data showing bright nuclear emission that is resolved into a triple source of 0.7″ extent with the central source apparently associated with the nucleus. Gallimore, Baum and O'Dea (2004) present VLBA images at 5 and 8.4 GHz of the nuclear radio source. The source S1, supposedly at the position of the hidden AGN, is resolved into 0.8 pc extended emission that is oriented almost perpendicularly to the radio jet axis, but closely aligns to a H_2O maser[10] disk.

---

10 Maser emission occurs when spectral lines (in the microwave region) of molecules are strongly amplified. Maser is an acronym for "Microwave Amplification by the Stimulated Emission of Radiation".

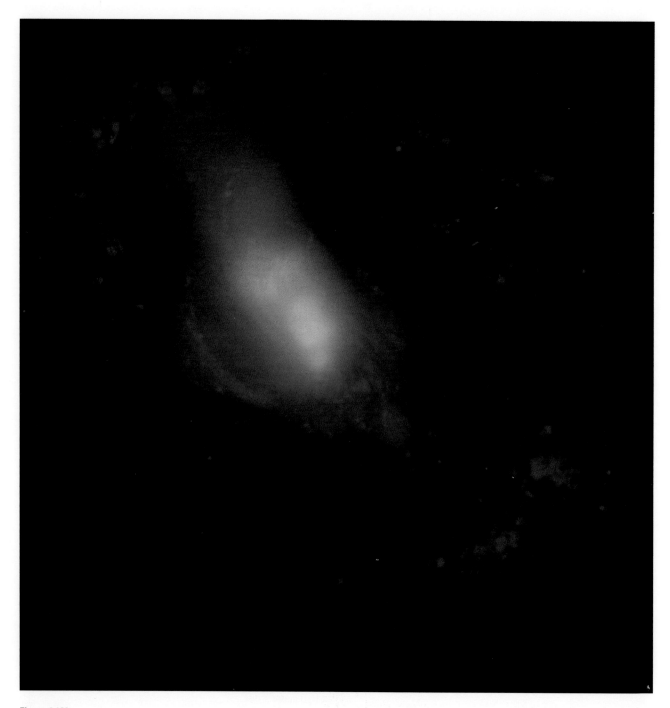

Figure 4.136

NGC 1068/M 77: Soft X-ray, 0.4–0.8 keV (green), 0.8–1.3 keV (blue), Optical (red),
36″ across.

Figure 4.137

NGC 1068/M 77: Optical, 1ʹ across.

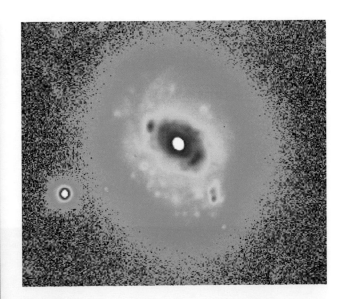

Figure 4.138

NGC 1068/M 77: Optical Hα + [N II].

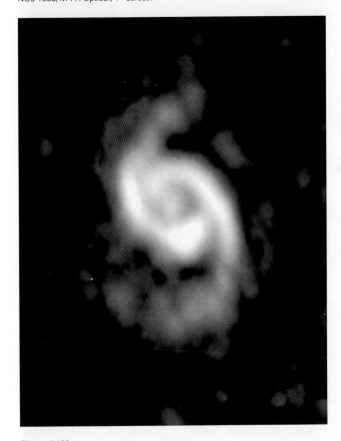

Figure 4.139

NGC 1068/M 77: Radio CO 2.6 mm, 110ʺ vertically.

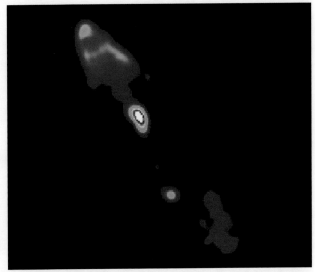

Figure 4.140

NGC 1068/M 77: Radio 6 cm continuum, ~15ʺ across.

4.5.2 NGC 1365

NGC 1365 is a Fornax Cluster member at 16.9 Mpc. It is an SBb(s)I type inclined at ~40°. It has an active nucleus of Seyfert 1.8 type.[11]

A multicolor optical image of NGC 1365 is shown in Figure 1.1.

An optical and X-ray image of NGC 1365 is shown in Figure 1.12.

---

11 Osterbrock (1981) proposed Seyfert 1.5, 1.8, and 1.9 subclasses, with numerically larger subclasses having weaker broad lines relative to narrow lines.

■ NGC 1365:

Turner, Urry and Mushotzky (1993) present ROSAT PSPC data showing discrete sources (probably X-ray binaries) close to the active nuclear source. Risaliti *et al.* (2005) use Chandra and XMM-Newton to detect X-ray spectral changes across several weeks. The authors suggest the variability is due to variation in the properties of a line-of-sight absorber, e.g. a circumnuclear torus.

Beck *et al.* (2005) present a radio continuum study with magnetic field strengths of 60 μG existing in the central star-forming regions,

decreasing to 20–30 μG along the bar. A decoupling of the magnetic field from the molecular clouds suggests that diffuse gas on kpc scales is controlled by the magnetic field. Jorsater and Van Moorsel (1995) used the VLA to observe the H I distribution which is similar to the optical, however with no evidence of a bar. H I velocities show a rapidly dropping rotation curve, well fitted by Keplerian motions in the outer parts.

Lindblad (1999) presents an extensive review on the kinematics and structure of the galaxy.

Figure 4.142
NGC 1365: Near-IR – central, ∼1.5′ across.

Figure 4.141   (opposite page)
NGC 1365: Far-UV 1400–1700 Å (blue), near-UV 1800–2750 Å (red), 15′ across.

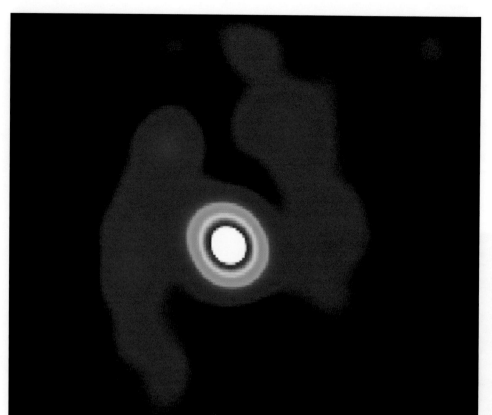

Figure 4.143
NGC 1365: Radio 4.8 GHz continuum,
30 ″ vertically.

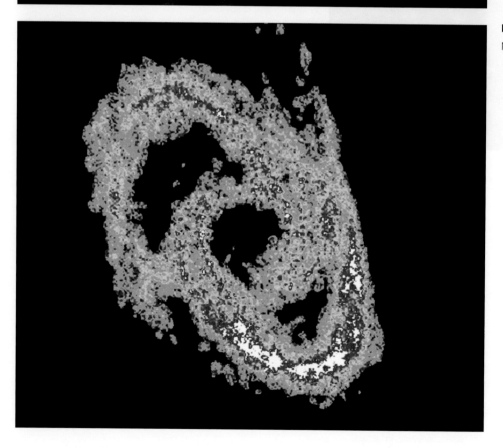

Figure 4.144
NGC 1365: Radio H I, 13.3 ′ across.

4.5.3 NGC 3031/Messier 81

NGC 3031 or Messier 81 is an Sb(r)I-II galaxy at 1.4 Mpc. It has been an important galaxy in the study of star formation and density-wave theory. It has an inclination of 58°.

Optical and H I observations (Yun, Ho and Lo 1994) of NGC 3031/M 81 and NGC 3034/M 82 (the dominant members of the M 81 Group) are shown in Figure 4.120.

NGC 3031/M 81 is given a secondary classification of I.

■ NGC 3031/M 81:

Swartz *et al.* (2002) use CXO ACIS observations to detect nine luminous soft X-ray sources, most of which can be explained as accreting white dwarfs powered by surface nuclear burning. The Hα line in the nucleus has a full width of $\sim$7000 km s^{-1}, and its luminosity is $\sim$20 times smaller than the lowest luminosity classical Sy 1 NGC 4051. The ionization level is low (compared to Seyfert nuclei) and the nucleus is probably best described as a LINER (Ho, Filippenko and Sargent 1996).

Kendall *et al.* (2008) derive the density wave structure via residual mass maps using Spitzer, optical and 2MASS data. The offset between the density wave and gas shocks (seen via 8 μm dust emission) implies a long-lived spiral structure. Gordon *et al.* (2004) investigate star formation and dust content using Spitzer, UV, Hα and 20 cm continuum images.

H I observations (Yun, Ho and Lo 1994) of NGC 3031/M 81 and NGC 3034/M 82, the dominant members of the M 81 Group (Figure 4.120) show dramatic evidence of tidal interactions (filaments) between group members that is not seen in the optical.

Figure 4.145

NGC 3031/M 81: Far-UV 1400–1700 Å (blue), near-UV 1800–2750 Å (red),
40 ′ across.

Figure 4.146 (opposite page)
NGC 3031/M 81: Optical, F435W (blue), F606W (green), F814W (red) – central,
12.9 ′ across, N is ∼20° CCW from up.

Figure 4.147

NGC 3031/M 81: Mid-IR 24 μm, far-IR 70, 160 μm, N is 91° CW from up.

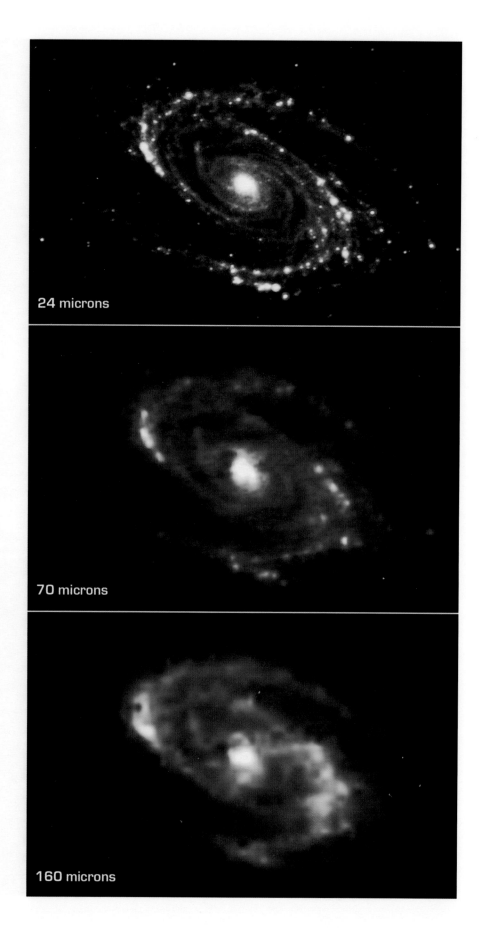

24 microns

70 microns

160 microns

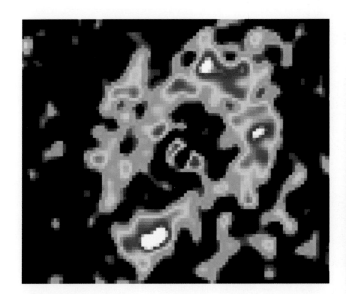

Figure 4.148

NGC 3031/M 81: Radio 2.8 cm continuum.

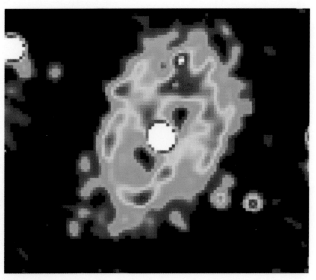

Figure 4.149

NGC 3031/M 81: Radio 1.49 GHz continuum, 16′ across.

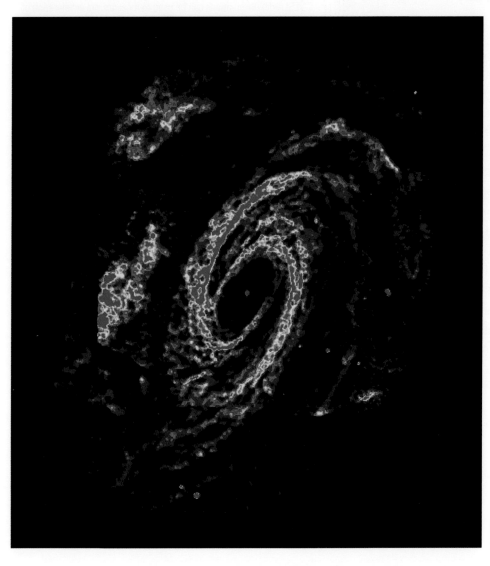

Figure 4.150

NGC 3031/M 81: Radio H I, 0.65° across.

4.5.4 NGC 4258/Messier 106

NGC 4258 or Messier 106 is an Sb(s)II galaxy at an inclination of ~70°. It is an active galaxy of LINER type at a distance of 6.8 Mpc. Two bright arms in its inner regions are strong in Hα and radio continuum.

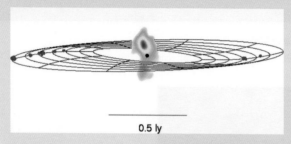

0.5 ly

Figure 4.151

The geometrical relationships of the jet emission (false color) in the nucleus, the disk of water molecules (red and blue dots) and the black hole (black dot) at the center of the galaxy NGC 4258/M 106.

Credit: Kindly provided by L. Greenhill.

Greenhill *et al.* (1995) present a remarkable VLBI map of the H_2O maser in the central 260 micro-arcseconds (μas) (or 0.009 pc). The masers describe a warped disk-like structure (Figure 4.151), displaying either solid body rotation or Keplerian motion. Models suggest a central mass of $2–4 \times 10^7 \, M_\odot$ (in a region of about half a light-year) and a mass density of $3.5 \times 10^9 \, M_\odot \, pc^{-3}$, indicating the presence of a supermassive object (e.g. a supermassive black hole) in the nucleus.

■ NGC 4258/M 106:

Pietsch *et al.* (1994) present a comprehensive X-ray analysis using ROSAT PSPC. They find extended emission from ~4×10^6 K gas in the halo. On the east side of NGC 4258 the X-ray emission is detected in the soft and hard band, on the west side only in the hard band. This can be explained by shadowing of the X-ray emission of the halo gas by cool gas in the disk. The active nucleus was not detected as an X-ray point source. Wilson, Yang and Cecil (2001) give a comprehensive description based on CXO observations and discuss "anomalous arms", that are unusually diffuse and amorphous. These were previously discovered through Hα imaging.

HST WFPC2, NICMOS and ground-based imaging by Siopis *et al.* (2009) is used to determine a SMBH mass of $3.3 \times 10^7 \, M_\odot$ in good agreement with maser dynamics derived estimates.

VLBI radio observations can be used to measure the acceleration of masers. Humphreys *et al.* (2008) find spiral structure in the nuclear accretion disk, and aim to determine a geometric maser-based distance to the galaxy to within ~3%.

Figure 4.152

NGC 4258/M 106: X-ray (blue), optical (gold), IR (red) and radio (purple),
9.2′ across. N is ∼20° CW from up.

Figure 4.153

NGC 4258: Far-UV 1400–1700 Å (blue), near-UV 1800–2750 Å (red), ~30′ across.

Figure 4.154

NGC 4258/M 106: Optical B, 6′ across.

Figure 4.155

NGC 4258/M 106: Radio CO 2.6 mm, 220″ across.

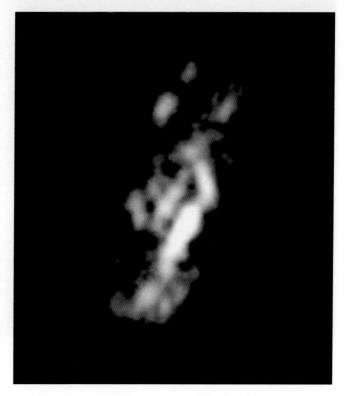

4.5.5 3C 273

3C 273 is a quasar at a redshift of 0.16. It was the first quasar formally identified (Schmidt 1963). The host galaxy is a blue, morphologically slightly disturbed elliptical galaxy. It has a jet south-west of the host galaxy starting at a radius of 10″ and extending 8″.

An X-ray, optical and infrared image of the jet in 3C 273 is shown in Figure 1.13.

An optical image of the 3C 273 host galaxy is shown in Figure 1.15.

Figure 4.156

3C 273: X-ray (0.1–10.0 keV). Field centered on the jet. 3C 273 at top, left. The image is 24″ across.

3C 273:

Marshall *et al.* (2001) use CXO and find the first knot in the jet (A1) is well fitted with a power law with $\alpha = 0.60 \pm 0.05$ (see power law in Section 1.5.5) and show a pure synchrotron model fits the spectrum of A1 over nine decades in energy. Jester *et al.* (2006) use deep CXO observations to support the synchrotron model for the jet. Espaillat *et al.* (2008) use XMM-Newton to detect a 3.3 ks quasi-period suggesting a mass of 7×10^6 $M_\odot$–8×10^7 $M_\odot$ for black hole last stable orbits between[12] 0.6 and 3.0 $R_{Schwarzschild}$ respectively. Reverberation mapping[13] estimates of black hole mass are higher, however, which may suggest the 3.3 ks period is a higher order oscillation of an accretion disk.

Hutchings and Neff (1991) have used CFHT to find a systematic optical color change along the jet that suggests an aging process. Jester *et al.* (2001) do not find any correlation between the optical flux and spectral index, as would be expected for relativistic electrons suffering strong cooling due to synchrotron emission.

The radio source shows superluminal motion with its apparent transverse velocity, $\beta_{app} \sim 8$ (Vermeulen and Cohen 1994), where

$$\beta_{app} = \mu z \frac{z}{H_0(1+z)} \left[\frac{1 + (1+2q_0z)^{1/2} + z}{1 + (1+2q_0z)^{1/2} + q_0z} \right]$$

for a Friedmann cosmology, where μ is its proper motion (in mas yr^{-1}), q_0 is the cosmological deceleration parameter, H_0 is the Hubble constant, and z, redshift.

Courvoisier (1998) presents a review article that discusses emission from radio to gamma rays. Emphasis is given to variability studies and the properties of the jet.

Figure 4.157

3C 273: Optical, F606W, $\sim 63''$ across, N is 15° CW from up.

---

12 Schwarzschild radius: the radius of the event horizon surrounding a non-rotating black hole. This quantity was first derived by Karl Schwarzschild in 1916.

13 Time delays between continuum and emission line variations are due to light travel time differences within the BLR. Hence emission lines "reverberate" to continuum changes. This was first postulated by Blandford and McKee (1982).

4.5.6 NGC 4486/Messier 87

NGC 4486 or Messier 87 is the central, dominant E0 galaxy, at 16.8 Mpc in the Virgo Cluster. The galaxy is centrally located in a strong X-ray emitting halo of 10^7 K gas, that is central to the cluster. At least 800 globular clusters form an overly dense (compared to other less luminous ellipticals) halo, prompting some to suggest that many of the globular clusters have been captured during galactic mergers. Henry Curtis discovered the optical jet near the nucleus in 1918. The optical jet was later found to be a strong synchrotron source, and was also observed in the radio and X-ray regions. The radio source is known as Virgo A.

■ NGC 4486/M 87:

Aharonian *et al.* (2003) report a detection above 730 GeV using the HEGRA system, suggesting the first TeV emission from a radio galaxy (rather than from a blazar). Abdo *et al.* (2009) detect the galaxy at >100 MeV with Fermi LAT from 10 months of survey data.

Wilson and Yang (2002) describe CXO ACIS imaging spectroscopy of the jet. The galaxy nucleus and all the knots seen at radio and optical wavelengths, out to knot C, are detected. The ratio of X-ray-to-radio, or optical flux declines with increasing distance from the nucleus. At least three knots are displaced from their radio/optical counterparts, being tens of parsecs closer to the nucleus at X-ray wavelengths. The X-ray spectra of the nucleus and knots are well described by power laws absorbed by cold gas. Forman *et al.* (2007) detect filaments and bubbles, and a hard-energy (3.5–7.5 keV) ring at 13 kpc radius suggestive of a shock related to the presence of a SMBH.

The HST FOC mid-UV image (Boksenberg *et al.* 1992; Maoz *et al.* 1996) shows the bright nucleus, and UV emission from the jet. In May, 2005 a "blob" of matter in the jet, named HST-1, became brighter than the nucleus (Madrid 2009).

HST WFPC2 narrowband Hα+[N II] images (Ford *et al.* 1994) detect a small disk of ionized gas in the nucleus. The disk is approximately elliptical in shape, and its minor axis is close to the projected position of the synchrotron jet.

Radial velocity measurements (Harms *et al.* 1994) at $r = \pm 0''.25$ along the disk major axis shows gas recession and approach velocities of 500 ± 50 km s^{-1}. A central mass of $\sim 2 \times 10^9$ M$_\odot$ is deduced, suggesting a mass-to-light ratio $(M/L)_I = 170$. The authors conclude that the disk of ionized gas is feeding a SMBH of $\sim 2 \times 10^9$ M$_\odot$. This idea was proposed nearly 20 years earlier by Wallace Sargent, Peter Young and collaborators, based on Hale 5 m telescope data (Sargent *et al.* 1978).

Very round isophotes (Goudfrooij *et al.* 1994a) are detected in the optical.

The FRI radio source (Virgo A, 3C 274; Perley (unpublished); Birkinshaw and Davies 1985) contains a kpc-scale radio jet (Biretta, Zhou and Owen 1995). Junor, Biretta and Livio (1999) show the jet at 43 GHz with milliarcsecond resolution using the VLBA, 13 VLA, and five European-based antennae. The jet has a wide (60°) opening angle, and is strongly collimated between 30 and 100 R$_{Schwarzschild}$. For comparison the last stable orbit around a non-rotating black hole is 6 R$_{Schwarzschild}$.

Biretta (1993) gives a review of the observational status of the jet at radio, optical, and X-ray frequencies.

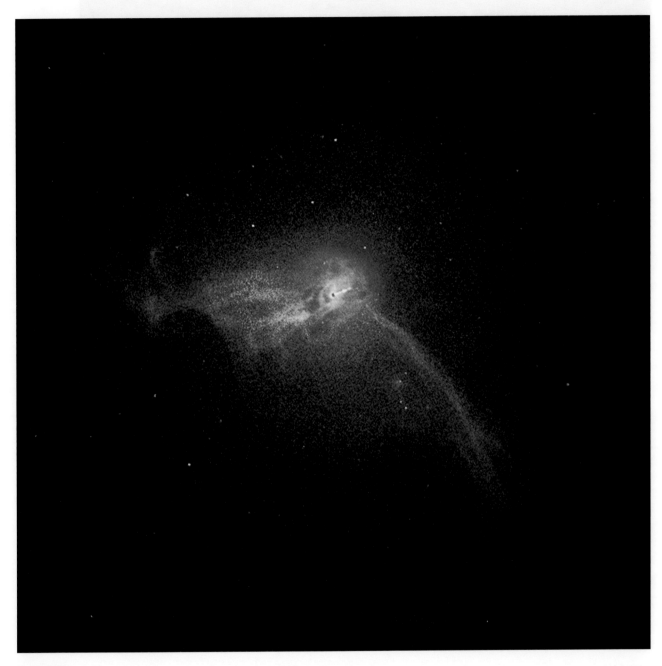

Figure 4.158
NGC 4486/M 87: X-ray, 11 ′ across.

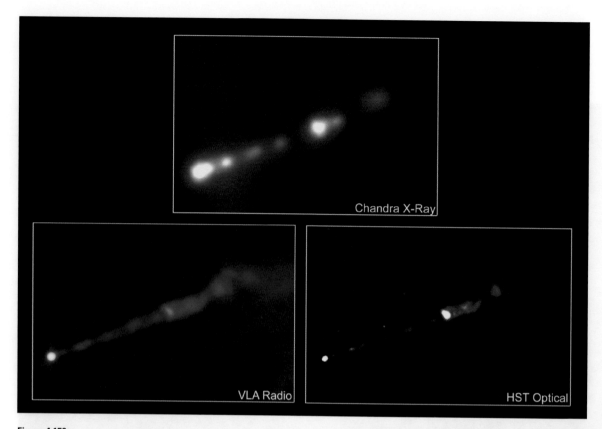

Figure 4.159

NGC 4486/M 87: X-ray + radio + optical –
jet, each panel 32″ across.

Figure 4.160

NGC 4486/M 87: Mid-UV 2300 Å.

Figure 4.161

NGC 4486/M 87: Mid-UV F220W, F250W, 4.8″ across. Nucleus is middle,
left, HST-1 is centered.

Figure 4.162

NGC 4486/M 87: Optical, 10′ across.

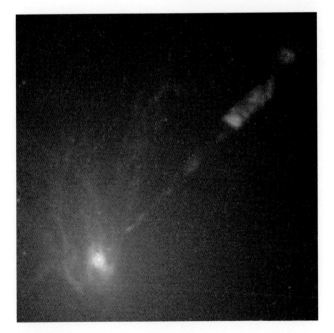

Figure 4.163

NGC 4486/M 87: Optical Hα — jet. Nucleus is at bottom, left. N is ~40° CCW from up.

Figure 4.164

NGC 4486/M 87: Near-IR 1.2 μm (blue), 1.6 μm (green), 2.2 μm (red), 6.3′ across.

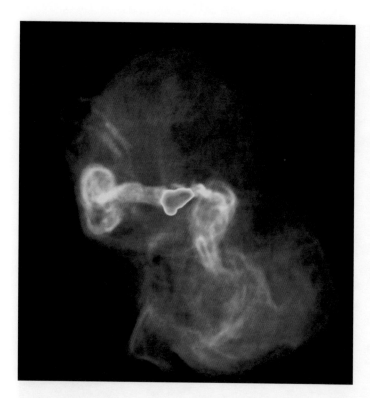

Figure 4.165

NGC 4486/M 87: Radio 90 cm continuum, 0.25° across.

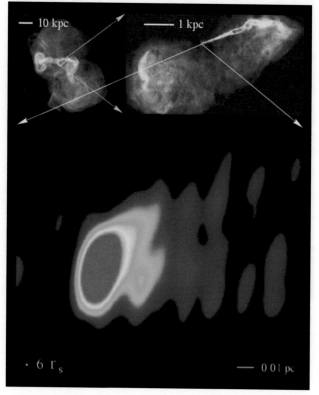

Figure 4.166

NGC 4486/M 87: Main image: Radio 43 GHz continuum, VLBA 4 milliarcseconds across. Top, left: 90 cm VLA; top, right: 20 cm VLA.

4.5.7 NGC 4594/Messier 104

The "Sombrero" Galaxy, NGC 4594 or Messier 104, is one of the most photogenic galaxies. At a distance of 20 Mpc it is an Sa^+/Sb^- type and has an inclination of ~ 80–$85°$.

It is dominated by an extensive dust lane and has a nearly edge-on disk and large nuclear bulge. Its nucleus is a compact and strong radio and X-ray source, and exhibits a LINER type spectrum (Heckman 1980).

Figure 4.169 was constructed by dividing the ESO VLT/FORS1 V image by itself smoothed with a two-dimensional Gaussian profile. This removes the uniform or smooth areas in the image and enhances the high spatial frequency features such as the dust bands. A large number of small, diffuse point-like sources, most of which are globular clusters, can be seen in the halo. The image was processed by Mark Neeser (Kapteyn Institute, Groningen) and Richard Hook (ST/ECF, Garching, Germany).

■ NGC 4594/M 104:

Fabbiano and Juda (1997) observed the galaxy with ROSAT HRI and detected a point-like nuclear X-ray source with L_X (0.1–2.4 keV) $\sim 3.5 \times 10^{40}$ ergs s^{-1} associated with the LINER nucleus. Pellegrini *et al.* (2002) combine 0.1–100 keV BeppoSAX and CXO ACIS observations to determine the spectral characteristics of the nuclear region. A dominant central X-ray point source coincides with the compact non-thermal radio source. The authors conclude that the LINER activity is linked to the presence of a low-luminosity AGN.

Krause, Wielebinski and Dumke (2006) use 4.86 GHz VLA and 8.35 GHz Heinrich Hertz Telescope continuum observations to detect the first large-scale magnetic field in an Sa type galaxy. The field is mostly parallel to the disk (some vertical components exist in the center and away from the mid-plane of the disk) and has an average value over the whole galaxy of 4 μG.

Figure 4.167

NGC 4594/M 104: X-ray (blue), optical (green), IR (red), 8.4′ across.

Figure 4.168

NGC 4594/M 104: Optical V (blue), R (green), I (red), 7′ across.

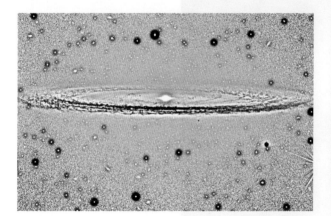

Figure 4.169

NGC 4594/M 104: Optical smoothed V, model subtracted.

Credit: P. Barthel, M. Neeser, ESO.

Figure 4.170

NGC 4594/M 104: Near-IR 1.2 μm (blue), 1.6 μm (green), 2.2 μm (red), 7.4′ across.

Figure 4.171

NGC 4594/M 104: Near-IR 3.6 μm (blue), 4.5 μm (green), 5.8 μm (orange) + mid-IR 8.0 μm (red), 9.6′ across, N is 15° CW from up.

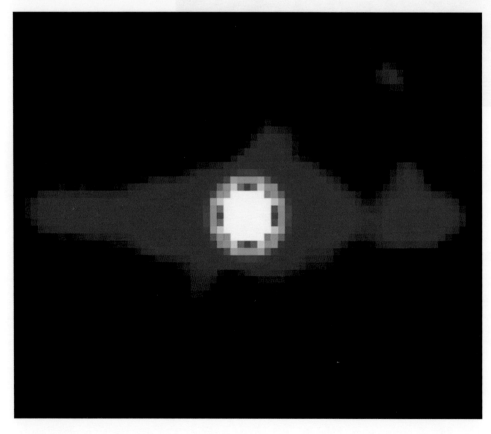

Figure 4.172

NGC 4594/M 104: Radio 6.2 cm continuum, 8′ across.

4.5.8 NGC 5128

An S0+S pec galaxy, NGC 5128 (associated with radio source Centaurus A) is the closest of all radio galaxies at 4.9 Mpc. It resides in the NGC 5128 Group. It possesses a prominent, warped dust lane, rich in atomic and molecular gas and luminous young stars. It is often considered to be the prototype Fanaroff–Riley Class I "low-luminosity" radio galaxy. Its globular cluster population possesses a bimodal metallicity distribution. Deep optical images reveal faint major axis extensions as well as a system of filaments and shells suggesting that NGC 5128 has experienced a major merging event at least once in its past. The galaxy has a very compact, subparsec nucleus exhibiting intensity variations at radio and X-ray wavelengths. The galaxy has jets in the optical, radio and X-ray at a variety of spatial scales.

An X-ray, optical and submillimeter image of Centaurus A is shown in Figure 1.7.

A secondary classification of I is adopted.

■ NGC 5128:

Aharonian *et al.* (2009) have detected >100 GeV gamma-ray emission from NGC 5128/Cen A using the High Energy Stereoscopic System (HESS). Over 120 hours, no significant variability was seen, although the detectable time-scales and variability strengths were limited.

Döbereiner *et al.* (1996) present ROSAT HRI observations that show X-ray emission from the nucleus, jet, halo and discrete sources. Emission from a region near the south-western inner radio lobe was discovered. Kraft *et al.* (2000) present CXO HRC observations that resolve individual knots of emission in the inner jet and also suggest the presence of an X-ray counter-jet. Karovska *et al.* (2002) detect arc-like structures in CXO imaging data.

Emission line filaments and young star associations have been detected in the middle radio lobe, suggesting radio (source) induced star formation. Morganti *et al.* (1991) show that optical line emitting filaments can be explained by excitation from an anisotropic[14] nuclear radiation field, suggesting that NGC 5128 is a blazar (see Section 1.5.5) with beamed radiation directed towards the optical jet. Marconi *et al.* (2001) measure the nuclear velocity field and detect a nuclear disk of ionized gas in the central $2''$ displaying Keplerian rotation. The authors suggest the existence of a central SMBH of $2.0 (+3.0, -1.4) \times 10^8$ $M_\odot$. Rejkuba *et al.* (2005) use deep HST ACS images to detect the horizontal branch (red clump) for the first time in an E/S0 galaxy. A wide red giant branch and asymptotic giant branch are visible, but not blue horizontal-branch stars.

The radio source extends to a diameter of $10°$, and contains outer radio lobes, a middle lobe ($23'$ from the nucleus), inner lobes ($12'$ in extent) and an inner jet of projected size 50 milli-arcseconds or 1 pc. The direction of the radio jet emission is at right angles to the large equatorial dust lane. The VLBI 8.4 GHz image (Jones *et al.* 1996) shows the core of the source (brightest feature in image), as well as the sub-pc scale radio jet to the north-east of the core and the sub-pc scale counter-jet to the south-west of the core. The VLBI contour maps at 4.8 GHz and 8.4 GHz (Jauncey *et al.* 1995; Figure 4.181) show the core of the radio source. Since the component which is brightest at 8.4 GHz and faintest at 4.8 GHz is the most active component it is most likely to be the active nucleus. Time series contour maps (Tingay *et al.* 1996, 1998) show VLBI images at 8.4 GHz taken between 1991 and 1995. The maps show that the apparent speed of components within the sub-pc scale jet are subluminal with velocities of $\sim 0.12 \pm 0.03c$ (and a slower component of $0.01 \pm 0.03c$).

H I observations (van Gorkom *et al.* 1990) show cold gas associated with the warped dust lane. H I velocities in the outer parts do not follow the inner kinematic pattern, and add support to a scenario that the dust lane is a direct result of a recent merger event.

Israel (1998) presents a multiwavelength review paper on NGC 5128.

---

14 An isotropic distribution is the same in all directions. In this case anisotropy suggests a preferential direction of radiation.

Figure 4.173

NGC 5128: X-ray (blue) + optical (yellow-orange) + radio (green-pink), 18′ across.

Figure 4.174

NGC 5128: X-ray (0.1–10.0 keV), 18′ across.

Figure 4.175

NGC 5128: Far-UV (blue), near-UV (red).

Figure 4.176

NGC 5128: Optical – greyscale, 17′ across.

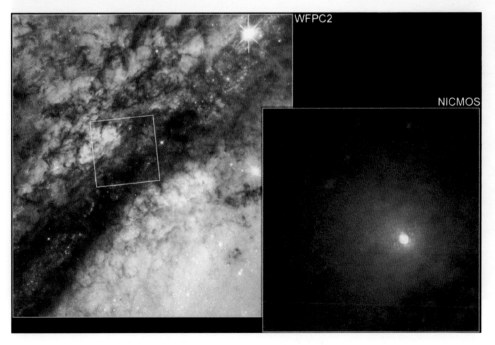

Figure 4.177

NGC 5128: Optical (left); near-IR (right), 1.87 μm (Pα) – central.

Figure 4.178

NGC 5128: Near-IR 1.2 μm (blue), 1.6 μm (green), 2.2 μm (red), 7.8′ across.

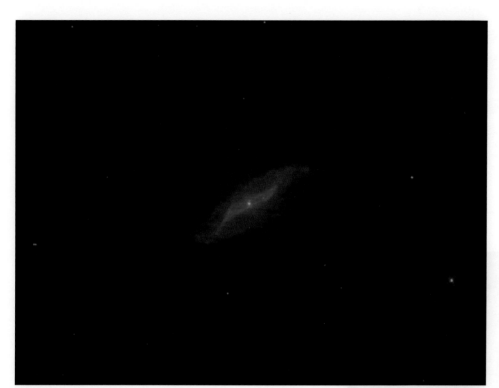

Figure 4.179

NGC 5128: Near-IR 5.8 μm + mid-IR 8.0 μm (both red), 27.1 ′ across.

ngc5128_VLBI_8.4GHz

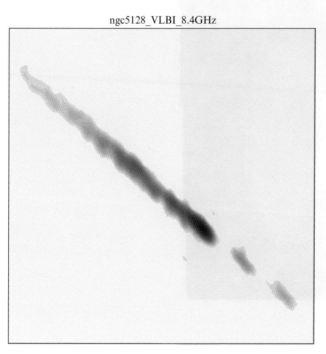

Figure 4.180

NGC 5128: Radio 8.4 GHz continuum – jet, a counter-jet is to the bottom, right of the bright source. The field is 100 milliarcseconds across.

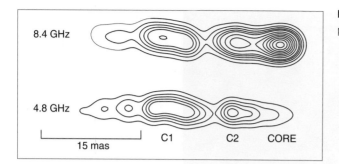

Figure 4.181

NGC 5128: Radio 4.8 and 8.4 GHz VLBI contour – jet.

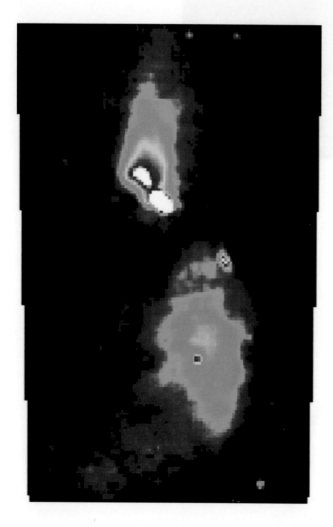

Figure 4.182

NGC 5128: Radio 6.3 cm continuum, 4° across.

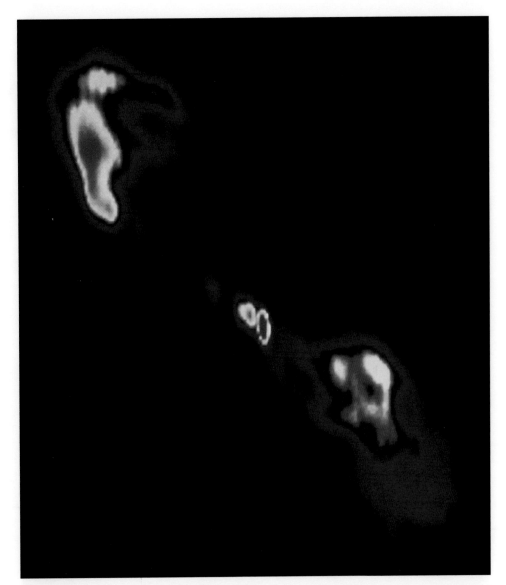

Figure 4.183

NGC 5128: Radio 1.4 GHz, 11′ across.

4.5.9 A1795 #1

A1795 #1 is the centrally dominant, early type, brightest galaxy in the rich cluster Abell 1795 at 243.9 Mpc. It has a redshift of 0.063 and is associated with the 4C 26.42 radio source of type FRI.

A secondary classification of I is adopted.

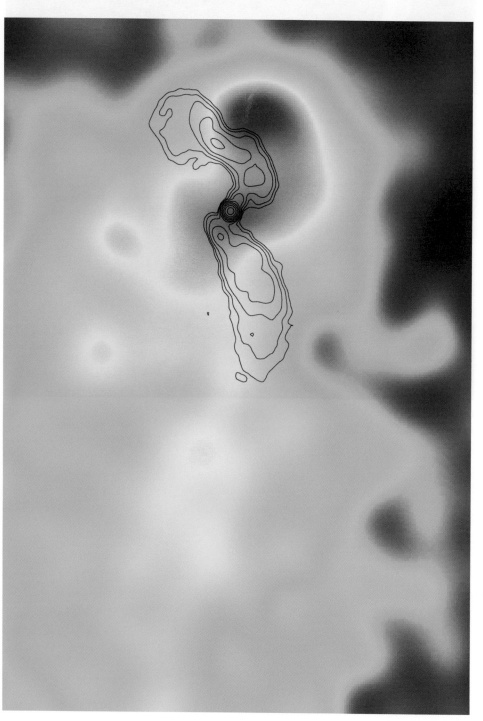

Figure 4.184

A1795 #1: X-ray (0.1–10.0 keV), radio 3.6 cm (contour), 18″ across.

A1795 #1:

The cluster is a strong source of X-rays, and the central region contains cooling gas at a rate of $\sim$300 $M_\odot$ yr^{-1} (Edge, Stuart and Fabian 1992). Briel and Henry (1996) present ROSAT PSPC X-ray temperature maps that clearly show a central cooling gas region, whilst outside of this the temperature appears constant. Fabian *et al.* (2001) use CXO data to report the discovery of a 40″ long X-ray filament in the core of the cluster. The X-ray feature coincides with a filament of Hα + [N II] found previously by Cowie *et al.* and resolved in U into filaments by McNamara *et al.* (1996a) (see below). A1795 #1 at the head of the X-ray filament is probably moving through or oscillating in the cluster core.

McNamara *et al.* (1996a) present deep optical U imagery that shows remarkable filamentary-like structures within the halo of A1795 #1. Some of these correspond to warm gas emission filaments, whilst others could be stripped material from small galaxies that are gravitationally interacting with A1795 #1. McNamara *et al.* (1996b) detect low amounts (<7%) of optical polarization across a central blue lobe structure that is aligned with the radio lobes, suggesting that the lobes are not scattered light from an active nucleus but sites of star formation. Pinkney *et al.* (1996) present WFPC2 imagery that shows an extensive dust lane that coincides with Hα to the north-north-west. Strong Hα emission is present in the nucleus, and in a filament that extends 80 kpc to the south.

Liuzzo *et al.* (2009) use VLBA to image the associated radio source 4C 26.42 at 1.6, 5.0, 8.4 and 22.0 GHz. The source is two-sided, but proper motion limits of the parsec-scale jets suggest they are subrelativistic.

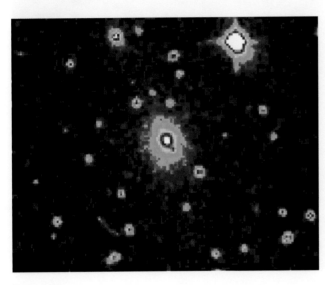

Figure 4.185

A1795 #1: Optical, 3′ across.

Figure 4.186

A1795 #1: Optical, F555W (blue), F702W (red), F555W+F702W (green), 18″ across, N is 55° CCW from up.

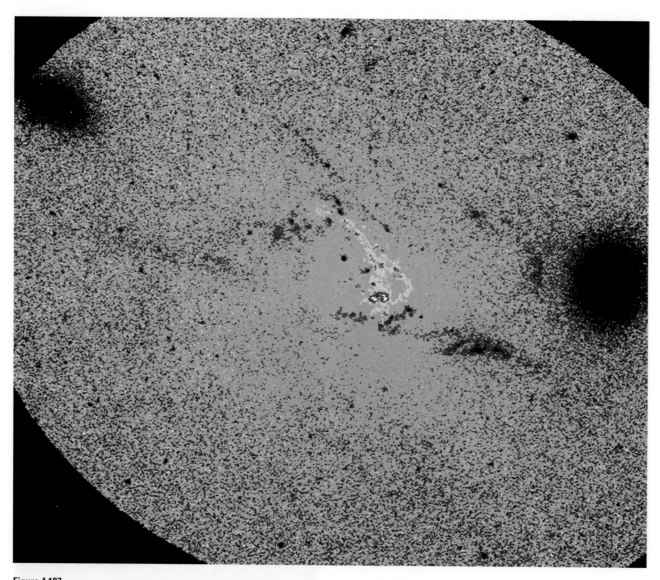

Figure 4.187

A1795 #1: Optical (V + R), model subtracted, 30″ across, N is 55° CCW from up.
Cluster galaxies (upper left and middle right) are also shown.

Figure 4.188   (opposite page)
Arp 220: X-ray (0.1–10.0 keV), 2′ across.

4.5.10 Arp 220

Arp 220 is an ultraluminous infrared galaxy ($L_{FIR} \sim 10^{12}$ $L_\odot$) that exhibits Sy 2 and starburst properties (Norris 1985; Smith, Aitken and Roche 1989). It is 71.3 Mpc distant. Secondary classifications of S and I are adopted.

■ Arp 220:

A central dust lane and the detection of double tidal tails (Sanders *et al.* 1988) suggest that Arp 220 has undergone a recent, near equal mass, disk–disk merger, and that this could have initiated either the AGN or starburst.

Shaya *et al.* (1994) detect many bright, compact objects in HST PC optical images that are close to the central radio continuum sources. Difficulties exist in determining the correct geometry and amount of dust obscuration, however many of these objects could be young stellar systems. If so, several will have short ($\sim 10^7$ yr) dynamical[15] friction time-scales and will rapidly merge into the nucleus.

Two nuclei are separated by $\sim 1''$ in the near-IR (Norris 1985).

Mid-IR spectral observations (Smith, Aitken and Roche 1989) suggest that a heavily obscured Seyfert nucleus is present and starburst regions exist outside the high extinction areas.

Norris (1988) presents VLA continuum images that show the presence of two nuclei separated by $\sim$350 pc. Lonsdale *et al.* (2006) use VLBI at 18 cm with 1 pc spatial resolution to detect 20 (east nucleus) and 29 (west nucleus) unresolved sources. The appearance of several sources on short time-scales strongly suggests these are radio supernovae with a birth rate of 4 ± 2 yr^{-1}.

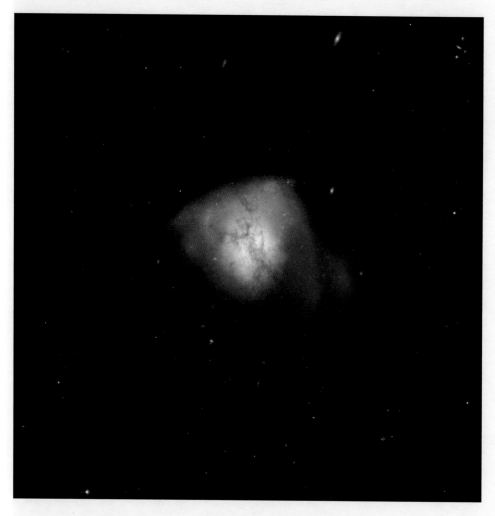

Figure 4.189

Arp 220: Optical, $3'$ across, N is 20 degrees CW from up.

---

15 As an object moves through other stars, gas and dark matter, it experiences a drag. This is referred to as dynamical friction and it is due to the object accelerating other objects in its wake, thus increasing their energy. As a consequence of the conservation of energy, the object loses energy and slows down. Over time, this causes the object, in this case, to spiral in towards the center of the host galaxy.

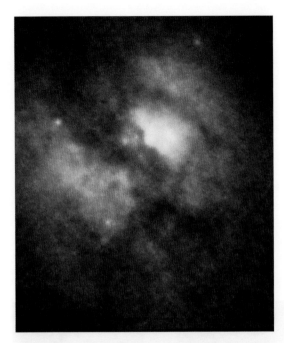

Figure 4.190

Arp 220: Optical – nucleus, 20″ across, N is 6° CCW from up.

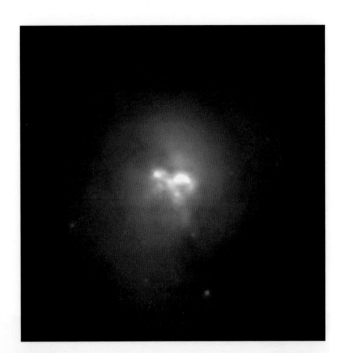

Figure 4.191

Arp 220: Near-IR – nucleus, 11″ across.

Figure 4.192

Arp 220: Radio 4.9 GHz continuum, 4″ across.

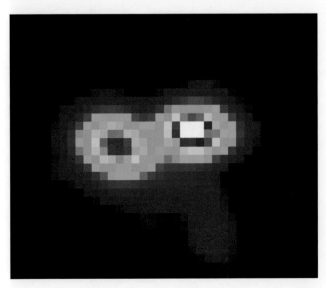

4.5.11 Cygnus A

The archetype of powerful radio galaxies, Cygnus A (3C 405) is associated with an early type galaxy with a redshift of 0.057, indicating a distance of 221.4 Mpc. It was initially detected at 160 MHz by Grote Reber in 1939. Cygnus A has a double structure (type FRII) that extends ~120 kpc. It has gained interest in light of the unified scheme of AGN, specifically, whether it hosts a "buried" or hidden quasar.

Figure 4.193

Cygnus A: X-ray (0.1–10.4 keV), 3.3′ across.

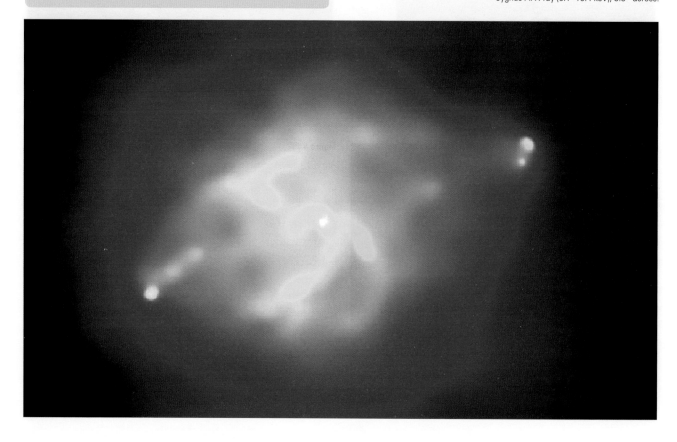

■ Cygnus A:

ROSAT HRI X-ray observations (Carilli, Perley and Harris 1994) show dramatic indications of modification of the hot, 10^7 K cluster gas by the jets and lobes of the radio source. Deficits in X-ray emission suggest that the hot ISM is removed or excluded by the passage of the radio jets and lobes.

Wilson, Young and Shopbell (2000) describe CXO observations of the radio (lobe) hot-spots. All four hot-spots were detected in X-rays with a very similar morphology to their radio structure. X-ray spectra have been obtained for the two brighter hot-spots (A and D) and both are well described by a power law with photon index $\gamma = 1.8 \pm 0.2$. Whilst thermal X-ray models are ruled out, the authors suggest that the observations strongly support synchrotron self-Compton[16] models of the X-ray emission, as proposed by Harris, Carilli and Perley on the basis of ROSAT observations.

Young *et al.* (2002) discuss the properties of the active nucleus based on CXO ACIS and Rossi X-Ray Timing Explorer (RXTE) observations. The hard (>2 keV) X-ray emission seen by CXO is spatially unresolved with a size <1″ (1.5 kpc) and coincides with the radio and near-infrared nuclei. In contrast, the soft (<2 keV) emission exhibits a bipolar nebulosity that aligns with the optical continuum and emission-line structures. Also, the soft X-ray emission coincides with [O III] and Hα + [N II] nebulosity detected by HST. Narrow X-ray emission lines from highly ionized Ne and Si, as well as the Fe Kα line at 6.4 keV, are detected in the nucleus.

Steenbrugge, Blundell and Duffy (2008) use a deep, 200 ks CXO ACIS image to study the X-ray jets. A counter-jet is seen, offset from the radio jet suggesting it is older, consisting of inverse Compton scattered CMB photons, and is not related to the current radio jet. A 10^6 yr time-scale of jet activity is predicted.

The outer radio lobes possess hot spots at their extremities. The north-west radio lobe is connected to core emission by a highly collimated kpc-scale jet (Perley, Dreher and Cowan 1984), whilst a south-east kpc-scale jet has also been detected (Carilli 1989).

VLBI imaging (Carilli, Bartel and Linfield 1991) detects a nuclear pc-scale radio jet that is aligned to within 4° of the north-west kpc-scale jet, whilst a south-east pc-scale jet is not found.

A multiwavelength workshop proceedings *Cygnus A – Study of a Radio Galaxy* is edited by Carilli and Harris (1996).

---

16 Synchrotron self-Compton radiation is due to inverse-Compton scattering of synchrotron radiation by the relativistic electrons that produced the synchrotron radiation.

Figure 4.194

Cygnus A: Optical, 3.3′ across.

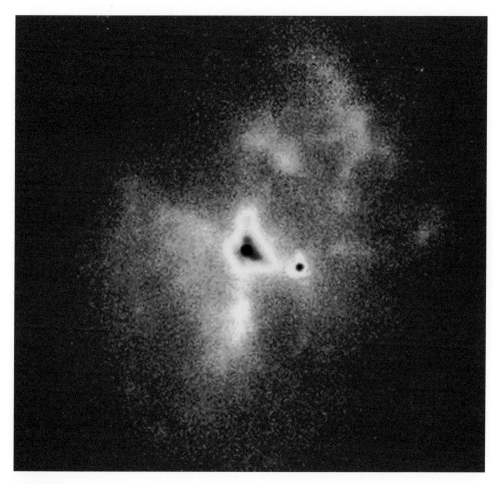

Figure 4.195

Cygnus A: Near-IR K′, ~1″ across.

Credit: Lawrence Livermore National Laboratory/W.M. Keck Observatory.

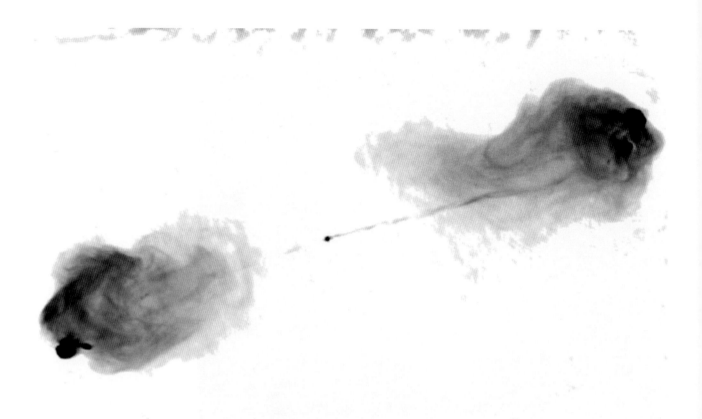

Figure 4.196

Cygnus A: Radio 5 GHz continuum VLA – greyscale.

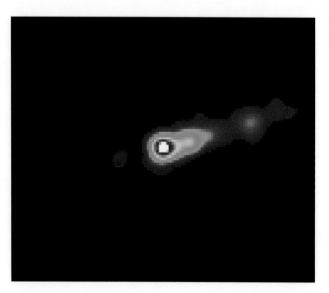

Figure 4.197

Cygnus A: Radio 5 GHz continuum VLBI, 40 milliarcseconds across.

Appendices and references

APPENDIX A
Telescopes and instruments

The following will briefly describe the main telescopes and instruments used to obtain the images presented in the atlas or those used in observations discussed in the text. Detectors are not described. Since there exists significant overlap between telescopes used for submillimeter and radio observations these are considered together.

A1 Gamma ray
$kT > 500$ keV

The Compton Gamma Ray Observatory (CGRO; Figure A.1) was launched into Earth orbit at 450 km altitude on April 5th, 1991 and re-entered on June 4th, 2000. CGRO contained instruments that could detect radiation with energies from 15 keV to 30 GeV and it was the second of NASA's "Great Observatories".

Figure A.1 Deployment of Compton Gamma Ray Observatory. Credit: NASA.

These instruments included the Energetic Gamma Ray Experiment Telescope (EGRET) that detected events between 20 MeV and 30 GeV with a positional accuracy of ~1°. The Imaging Compton Telescope (COMPTEL) covered 1–30 MeV with a positional accuracy of ~2°. Currently, gamma-ray imaging observations of nearby galaxies are restricted due to a small number of recorded events at poor positional accuracy.

The Fermi Gamma Ray Space Telescope (hereafter Fermi), formerly called the Gamma Ray Large Area Space Telescope or GLAST, was launched on June 11th, 2008 into a 560 km altitude orbit. It is detecting radiation between 8 keV and 300 GeV using the primary instrument, the Large Area Telescope (LAT), and the complementary GLAST Burst Monitor (GBM). The LAT has a large field of view, over 2 steradians (one-fifth of the entire sky), can measure the locations of bright sources to within 1 arcminute and is sensitive to photons from 30 MeV to greater than 300 GeV. The GBM cover X-rays and gamma rays between 8 keV and 30 MeV, overlapping with the LAT's lower energies.

A2 X-ray
Hard: $3 < kT < 500$ keV; soft: $0.1 < kT < 3$ keV

The High Energy Astrophysical Observatory-2 (HEAO-2), later named *Einstein*, was launched into low Earth orbit on November 13th, 1978 and operated until April, 1981. It was the first X-ray mission to use focusing optics with imaging detectors and produce angular resolution of a few arc seconds[1] ($''$) and a field of view of tens of arc minutes ($'$).

The Röntgen Observatory Satellite (ROSAT; Trümper 1983) was launched on June 1st, 1990 and is currently non-operational, though still orbiting the Earth. The ROSAT Position Sensitive Proportional Counter (PSPC) soft X-ray images spanning the energy range 0.1–2.4 keV, known as broadband, and 0.1–0.4 keV, known as softband, will be shown in this atlas. Typically, lower energy X-rays are referred to as "soft", while high-energy X-rays, not surprisingly, are designated "hard". ROSAT PSPC images have a minimum angular resolution of 25 $''$ (FWHM) over a field of view of 2°. The ROSAT High Resolution Imager (HRI) has an angular resolution of ~5 $''$ FWHM and a ~30 $'$ field of view.

The Chandra X-Ray Observatory (CXO), formerly the Advanced X-Ray Astrophysics Facility or AXAF), was the U.S. follow-on to *Einstein*, and was launched on July 23rd, 1999. Chandra was the third of NASA's "Great Observatories". Chandra has a highly elliptical orbit, 133,000 by 16,000 km, which puts it above the van Allen belts for ~85% of its orbit. CXO carries a set of nested high angular resolution mirrors, two imaging detectors (Advanced CCD Imaging Spectrometer, ACIS and High Resolution Camera, HRC), and two sets of transmission gratings (for spectral observations). Important CXO features are: an order of magnitude improvement in angular resolution (above ROSAT and *Einstein*), good sensitivity from 0.1 to 10 keV, and the capability for high spectral resolution observations over most of this range.

---

1 Seconds of arc are denoted as $''$; minutes of arc as $'$; degrees as $°$. See Appendix F, Table 7, page 237.

XMM-Newton was launched on December 10th, 1999, into a 48 hour elliptical orbit with an apogee of 114,000 km, and perigee of 7000 km. It has a spectral range of 0.1–12.0 keV (120–1 Å), and its primary imager is the European Photon Imaging Cameras (EPIC).

A3 Extreme ultraviolet
Extreme (EUV): 100–912

The Extreme Ultraviolet Explorer (EUVE) was launched on June 7th, 1992 and re-entered on January 30th, 2002. The EUVE and ROSAT Wide Field Camera (WFC) both had extreme UV (EUV) survey imaging capabilities. However, they do not possess the sensitivity nor angular resolution to provide images of galaxies for this atlas. They have been successfully used to produce catalogues of bright sources both of Galactic and extragalactic origin.

A4 Far- and mid-ultraviolet
Far-UV: 912–2000 Å; mid-UV: 2000–3300 Å

The Ultraviolet Imaging Telescope (UIT; Stecher *et al.* 1992; Figure A.2) was flown aboard the Space Shuttle *Astro-1*

Figure A.2 *Astro-1* instruments in the Space Shuttle Bay. UIT is the large cylinder blanketed with thermal insulation. Credit: STScI and NASA.

mission (December 2–11, 1990) and *Astro-2* mission (March 2–18, 1995). UIT was a 38 cm telescope allowing imaging with a 40′ field of view. Most images were made with either a broadband mid-UV (central wavelength λ_c ~2000–3000 Å) or far-UV (λ_c ~1500–1700 Å) filter. The filters used in atlas images are mid-UV A1 and A5 (λ_c ~2800 Å and ~2500 Å, respectively) and far-UV B1 and B5 (λ_c ~1500 Å and ~1600 Å, respectively).

The Swift telescope was launched into a low-Earth orbit on November 20th, 2004. It has a UV/Optical Telescope (UVOT) with wavelength coverage from 170 to 650 nm. Its prime mission is to detect gamma ray bursts (GRBs).

The Galaxy Evolution Explorer (GALEX) was launched on April 28th, 2003. It is a 50 cm Ritchey–Chretien telescope with far-UV (FUV: 1400–1700 Å) and mid-UV (NUV: 1800–2750 Å) imaging capability.

Ultraviolet (far- and mid-) observations have been carried out with the Hubble Space Telescope (HST), which was launched on April 24th, 1990 into a near circular low (560 km) Earth orbit. HST (Figure A.3) was the first of NASA's "Great Observatories". The HST Faint Object Camera (FOC; Paresce 1990) used F175W, F220W and F275W broadband filters with effective wavelengths (λ_{eff}) of ~1900 Å, ~2300 Å and ~2800 Å, respectively, with bandwidths ($\Delta\lambda$) of ~500 Å. FOC was removed from HST in March, 2002.

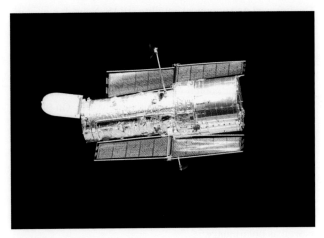

Figure A.3 The Hubble Space Telescope in orbit during the first servicing mission in 1993. Credit: NASA.

A5 Optical
3300 to 8000 Å

Optical images in the atlas have been taken with a variety of ground-based telescopes as listed in Appendix C (page 224). The instruments used were typically prime focus cameras with charge coupled devices (CCDs) as detectors. Hubble Space Telescope Wide Field Planetary Camera (WFPC; Westphal 1982), Wide Field Planetary Camera-2 (WFPC2;

Holtzman *et al.* 1995), Advanced Camera for Surveys (ACS; Ford *et al.* 1998) and Wide Field Camera 3 (WFC3) CCD images are also presented.

Standard broadband ($\Delta\lambda > 650$ Å) optical filters include U ($\lambda_c \sim 3600$ Å), B ($\lambda_c \sim 4400$ Å), V ($\lambda_c \sim 5500$ Å), R ($\lambda_c \sim 6500$ Å) and I ($\lambda_c \sim 8200$ Å). Narrowband ($\Delta\lambda < 100$ Å) filters are also used to isolate individual emission lines such as Hα (6563 Å) and [N II] (6548,6583 Å).

A6 Near-infrared
Near-IR: 0.8–7 μm

Near-IR images have been taken with the Steward Observatory 2.3 m and 1.6 m, University of Hawaii 88 inch and the KPNO 1.3 m telescopes. Standard near-IR filters include J ($\lambda_c \sim 1.25$ μm), H ($\lambda_c \sim 1.65$ μm), K ($\lambda_c \sim 2.2$ μm), L ($\lambda_c \sim 3.5$ μm) and M ($\lambda_c \sim 5$ μm). Mosaiced three-color near-IR images from the Two Micron All Sky Survey (2MASS) are also shown. 2MASS used two 1.3 m telescopes, one at Mt Hopkins, Arizona, and one at CTIO, Chile. Each telescope was equipped with a three-channel camera, each channel consisting of a 256×256 array of HgCdTe detectors, capable of observing the sky simultaneously at J, H, and K_s (2.17 μm).

A7 Mid- and far-infrared
Mid-IR: 7–25 μm; far-IR: 25–300 μm

The Infrared Astronomical Satellite (IRAS; Neugebauer *et al.* 1984) was launched on January 25th, 1983 into a 900 km altitude orbit. It produced an all-sky survey covering 96% of the sky at wavelengths of 12, 25, 60 and 100 μm. IRAS had a 0.57 m diameter primary mirror and the telescope was mounted in a liquid helium cooled cryostat. The mirrors were made of beryllium and cooled to approximately 4 K. On November 22nd, 1983 the survey finished due to the depletion of on-board liquid helium.

The Infrared Space Observatory (ISO; Kessler *et al.* 1996) was launched on November 17th, 1995 and ceased observation in April, 1998. ISO operated from 2.5 to 240 μm. It had a 60 cm diameter primary mirror and several instruments including ISOPHOT (Lemke *et al.* 1996) that could perform imaging as well as polarimetry.[2]

The Spitzer Space Telescope (SST; formerly SIRTF, the Space Infrared Telescope Facility; Werner *et al.* 2004) was launched on August 25th, 2003 and is the fourth and final telescope in NASA's family of "Great Observatories". SST is in an Earth trailing heliocentric orbit and moves away from Earth at ~ 0.1 AU[3] per year. SST has an 85 cm diameter mirror

---

2 Electromagnetic waves may travel in a preferred plane – unpolarized light does not have a preferred plane of vibration. Polarimetry is the measure of a preferred plane of propagation and the amount is called the polarization and can be between 0 and 100%.

3 The astronomical unit, or AU, is the average distance between the Earth and Sun.

and three cryogenically cooled science instruments capable of performing imaging and spectroscopy between 3.6 and 160 μm. Wide-field, broadband imaging is done by the Infrared Array Camera (IRAC) and the Multiband Imaging Photometer for Spitzer (MIPS).

A8 Submillimeter and radio
Submillimeter: 300 μm to 1 mm; radio: 1 mm and longer wavelengths

The 15 m diameter, alt-azimuth (alt-az) mounted James Clerk Maxwell Telescope (JCMT; Figure A.4) on Mauna Kea, Hawaii, operates specifically in the mm (radio) and submillimeter regions. Since, May 1997 the Submillimeter Common-User Bolometer Array (SCUBA) instrument on JCMT has produced observations between 350 and 850 μm. A new generation SCUBA-2 is expected to begin operation post-2009.

Figure A.4 A rare view of the James Clerk Maxwell Telescope after the telescope membrane had been removed in preparation for removing the secondary mirror unit. Credit: James Clerk Maxwell Telescope, Mauna Kea Observatory, Hawaii.

The Owens Valley Radio Observatory (OVRO), located near Bishop, CA, U.S.A., has a Millimeter Wavelength Array comprising six 10.4 m telescopes with half power beam widths (HPBW; a spatial resolution measure similar to FWHM) equal to 65 ″. The array is used for aperture synthesis mapping of millimeter line and continuum emission in the 2.7 and 1.3 mm windows. The Columbia Southern Millimeter-Wave Telescope is a 1.2 m diameter Cassegrain telescope located at CTIO, Chile and is described by Bronfman *et al.* (1988).

The Very Large Array (VLA), a 27-antenna radio telescope array, is located in Socorro, New Mexico, U.S.A. The telescope consists of 25 m diameter parabolic dishes, which can be placed along a Y-shaped pattern with each of the arms being 20 km long. The VLA can observe radiation between 1.3 and 22 cm and achieve an angular resolution similar to a telescope of size 27 km in diameter via aperture synthesis.

The Westerbork Synthesis Radio Telescope (WSRT) is a 3 km long array of fourteen 25 m antennas, located near Hooghalen, Netherlands. The Jodrell Bank 76 m radio telescope (now known as the Lovell Telescope) is an alt-az mounted telescope in Cheshire, England. The Effelsberg (near Bonn, Germany) 100 m radio telescope was the world's largest moveable radio telescope until August, 2000 when the 100 m × 110 m Robert C. Byrd Green Bank Telescope in Virginia saw "first light".

The Australia Telescope Compact Array (ATCA) in Narrabri, New South Wales, consists of six 22 m antennas (five antenna are located on a 3 km east–west railway and one a further 3 km to the west). The Parkes (New South Wales, Australia) radio telescope (Figure A.5) is a single 64 m diameter antenna on an alt-az mounting that operates between wavelengths of 1 and 70 cm (21–0.5 GHz).

Very long baseline interferometry (VLBI) is conducted by many radio antennas separated by large distances to achieve very high angular resolution. The signals from astronomical sources are recorded on large capacity disk drives along with accurate timing information (usually via a highly stable hydrogen maser clock) and processed by a signal correlator. The correlator removes known geometric delay and Doppler shift due to the Earth-based motion of the antennas. The Southern Hemisphere VLBI Experiment (SHEVE) array was an ad hoc array of radio telescopes in the Southern Hemisphere (mainly in Australia with occasional contributions from South Africa). The Australia Telescope National Facility (ATNF) now operates the Long Baseline Array (LBA) consisting of ATCA, Mopra and Parkes. There is an Australian VLBI National

Figure A.5 Parkes 64 m radio telescope. © Seth Shostak.

Facility, comprising the LBA plus University of Tasmania antennas at Hobart and Ceduna, and the Tidbinbilla antenna.

APPENDIX B
Source list of all-sky Galaxy images

The COMPTEL gamma-ray image was provided by R. Diehl, U. Oberlack and the COMPTEL team. References – Oberlack, U. *et al.* 1996, Astron. Ap. Suppl., **120**, C311; Diehl, R. *et al.* 1996, Astron. Ap. Suppl., **120**, C321; Oberlack, U. Ph.D. Thesis, MPE 1997.

The EGRET gamma-ray, soft X-ray, near-IR, mid- and far-IR, molecular hydrogen and atomic hydrogen images were compiled by the Astrophysics Data Facility (ADF) at Goddard Space Flight Center (GSFC). See "Multiwavelength Milky Way" at http://mwmw.gsfc.nasa.gov/ References – Gamma ray: Hunter, S.D. *et al.* 1997, Ap. J., **481**, 205; Thompson, D.J. *et al.* 1996, Ap. J. Suppl., **107**, 227. – Soft X-ray: Snowden, S.L. *et al.* 1995, Ap. J., **454**, 643. – Near-IR: Hauser, M.G.,

Kelsall, T., Leisawitz, D. and Weiland, J. 1995, COBE DIRBE Explanatory Supplement, Vers. 2.0, COBE Ref. Pub. No. 95-A (NASA/GSFC, Greenbelt, MD) – Mid- and far-IR: Wheelock, S.L. *et al.* 1994, IRAS Sky Survey Atlas Explanatory Suppl., JPL Publ. 94-11 (JPL, Pasadena) – Molecular hydrogen: Dame, T.M. *et al.* 1987, Ap. J., **322**, 706; Digel, S.W. and Dame, T.M. 1995, unpubl. survey update. – Atomic hydrogen: Burton, W.B. 1985, Astron. Ap. Suppl., **62**, 365; Hartmann, Dap and Burton, W.B. 1997, *Atlas of Galactic Neutral Hydrogen* (Cambridge University Press, Cambridge); Kerr, F.J. *et al.* 1986, Astron. Ap. Suppl., **66**, 373.

The Galaxy: Optical – Kindly provided by Prof. L. Lindegren, Director, Lund Observatory.

The Galaxy: Optical – Hα at 6563 Å – Kindly supplied by the Southern H-Alpha Sky Survey Atlas (SHASSA), "A robotic wide-angle Hα survey of the southern sky" by Gaustad, J.E., McCullough, P.R., Rosing, W. and Van Buren, D. 2001, P.A.S.P., **113**, 1326.

The 73 cm/408 MHz image is described by Haslam, C.G.T. *et al.* 1982, Astron. Ap. Suppl., **47**, 1 and was extracted from the NRAO CD-ROM "Images from the Radio Universe".

APPENDIX C
Source list of galaxy images (Table 4)

Galaxy/region	Tel./Instr.	Observ.	λ/ν/Energy/Filt.	Exp.	Res.	Image source
NGC 224/M 31						
X-ray	XMM-Newton/ EPIC MOS		0.3–12.0 keV	34.8+12.2 ks	∼6″	ESA/XMM-Newton+ STr
X-ray	CXO/ACIS		0.1–4.0 keV	54 h	∼2″	NASA/UMass/ ZL/QDW
{Far-UV+ {Mid-UV	Swift/UVOT		UVW1, UVM2,UVW2	∼24 h	2″	NASA/Swift/SI/EG
Opt.	HST/WFPC2		F160BW*,F300W, F555W,F814W	3.7 h		NASA/ESA/TL/ NOAO/AURA/NSF
Opt.	KPNO BS	Guhathakurta, Raychaudhury	B	22 × 500 s	∼3.5″	PG/PC/SR
Near-IR	2MASS		J,H,K_s	7.8 s	2″	2MASS
Mid-IR	Spitzer/IRAC	Bamby	8 μm	106 s	1.9″	NASA/JPL/PBa
Far-IR	IRAS		60 μm		†62″× 41″	IPAC
Far-IR	ISO/ISOPHOT		175 μm		112″	MH
Radio	IRAM		2.6 mm			MPIfR/IRAM/ CN/NN/MG+
Radio	Effelsberg	Beck	6 cm			MPIfR/RB/EMB/PH
Radio	VLA+ Eff. 100m	Beck, Berkhuijsen	20 cm	7 × ∼8 h (VLA) ∼12 h (Eff.)	45″	RB
Radio	WSRT	Brinks	21 cm line	24 × 12 h	24″× 36″	EB
SMC						
Soft X-ray	ROSAT/PSPC	Pietsch	0.1–2.4 keV	var.	≥30″	PK
{Near-IR+ {Mid-IR+ {Far-IR	Spitzer/ IRAC/ MIPS		3.6, 8.0,24.0, 70.0,160.0 μm			NASA/JPL-Caltech/ STScI
Radio	Columb. 1.2m	Rubio+	CO ($J = 1 \rightarrow 0$)		20″	MR
Radio	P64	Haynes+	1.4 GHz		15′	UK/RH

Table 4 (*cont.*)

Galaxy/region	Tel./Instr.	Observ.	λ/ν/Energy/Filt.	Exp.	Res.	Image source
NGC 300						
X-ray	XMM-Newton	Carpano+	0.3–1.0,1.0–2.0, 2.0–10.0 keV	66 ks	6″	SC+/ESA
{Far-UV+ {Opt.	GALEX		1400–1700 Å		4.2″	NASA/JPL-Caltech/ OCIW
Opt.	HST/ACS	Dalcanton	F450W,F555W, F814W			NASA/ESA/JD/BWi
Opt.	CTIO CS	Mackie	Hα+[N II]	1800 s	~4″	GHM
{Near-IR+ {Mid-IR	Spitzer		3.6,4.5,5.8, 8.0 μm			NASA/JPL-Caltech/ GH
Radio	VLA (C/D)	Puche+	21 cm line		50″ × 50″	NRAO
NGC 598/M 33						
X-ray	XMM-Newton	Pietsch	0.2–1.0,1.0–2.0, 2.0–12.0 keV		20″	WP/MPE/ ESA
Far-UV	UIT		B5	431.0 s	3″	NDADS
{Far-UV+ {Near-UV+	GALEX+					NASA/JPL-Caltech
{Near-IR+ {Far-IR	Spitzer					
Opt.	McDonald 30	Bothun	U			GB
Opt.	HST/WFPC2		F336W,F375N, F487N,F502N, F555W,F656N, F658N,F673N, F814W,F953N			NASA/HHT+ DG/JH/JW
{Near-IR+ {Mid-IR	Spitzer		3.6,4.5, 8.0,24.0 μm			NASA/JPL-Caltech/ Univ. of Ariz.
Radio	VLA+WSRT	Duric	6 cm			ND
Radio	VLA (D,C/D)	Condon	20 cm/ 1.49 GHz		~0.9′	NRAO
Radio	VLA (B,CS)+ WSRT		21 cm line			NRAO/AUI
NGC 891						
Opt.	Pal. 1.5m/ WF-PFUEI	Rand, Kulkarni, Hester	Hα	7500 s	~2.4″	RR
Opt.	WIYN 3.5m	Howk	V			CH/BS/WIYN/ NOAO/NSF
Near-IR	2MASS		J,H,K$_s$	7.8 s	2″	2MASS
Radio	VLA		21 cm line			NRAO
Radio	VLA		cont. nr. 21 cm			NRAO

Table 4 (*cont.*)

Galaxy/region	Tel./Instr.	Observ.	λ/ν/Energy/Filt.	Exp.	Res.	Image source
NGC 1399						
X-ray	CXO/ ACIS	Irwin	0.1–10.0 keV	18 h	~2″	NASA/CXC/ UA/JI+
Soft X-ray	ROSAT/PSPC	Jones	0.1–2.4 keV	53.5 ks	≥25″	USRDA
Opt.						NASA/STScI
Radio	VLA (C)	Killeen	6 cm/ 4.9 GHz	2 h	3.2″ × 10″	NK
LMC						
Soft X-ray	ROSAT/PSPC	var.	0.5–2.0 keV	~5–30 ks	1-5′	SS
Opt.	CTIO/CS		[O III],Hα, [S II]			UM/CTIO MCELS/ NOAO/AURA/NSF
Opt.	HST/WFPC2		F336W,F555W, F656N,F673N, F814W			NASA/NW/ JM-A/RBa
Near-IR	2MASS		J,H,K$_S$	7.8 s	2″	2MASS
Near,Mid-IR	Spitzer		3.6,8.0,24.0 μm	43,43,60 s		NASA/JPL-Caltech/ MM+SAGE
Radio	Mopra		CO ($J = 1 \rightarrow 0$)			AH/TW/JO/JP/ EM/MAGMA
Radio	ATCA		1.4 GHz			AH/LSS/CSIRO
NGC 2915						
Opt.	CTIO 1.5m/ RFP	Marlowe, Schommer	Hα	1800 s	~2″	AM
Opt.	AAT PFC	Meurer, Hatzidimitriou	R	400 s	1.9″	HGM
{Radio+	ATCA+	Meurer	21 cm	32.7 h	45″	HGM
{Opt. {	AAT PFC	Meurer, Hatzidimitriou	B	350 s	1.9″	
Malin 2						
Opt.	KPNO 2.1m	Bothun	Hα		~2″	SM/GB
Opt.	MDM 1.3m	McGaugh	I	5 m	~2″	SM/GB
NGC 5457/M 101						
Soft X-ray	CXO/ACIS	Kuntz	0.45–1.0, 1.0–2.0 keV	11 d 10 h	2″	NASA/CXC/ JHU/KK+
{Far-UV+ {Near-UV	GALEX		1400–1700, 1800–2750 Å		4.2″	NASA/ JPL-Caltech
Opt.	HST/ACS+WFPC2+ KPNO+CFHT		F435W, F555W, F814W			NASA/ESA/KK/FB/JT/ JMo/YCC/STScI

Table 4 (*cont.*)

Galaxy/region	Tel./Instr.	Observ.	λ/ν/Energy/Filt.	Exp.	Res.	Image source
{Near-IR+	Spitzer	Gordon	3.6,	85,		NASA/JPL-Caltech/
{Mid-IR			8.0, 24.0 μm	85,200 s		KG
Radio	Eff. 100m	Beck	6 cm/4.8 GHz			RB
Radio	WSRT	Kamphuis	21 cm line	16×12 h	$16'' \times 13''$	JK

NGC 6822

Near-IR	2MASS		J,H,K$_S$	7.8 s	$2''$	2MASS
Far-IR	IRAS		60 μm		$^\dagger 62'' \times 41''$	IPAC
Radio	VLA (D,C/D)	Condon	20 cm/ 1.49 GHz		$\sim$0.9$'$	NRAO

NGC 4406/M 86

{X-ray+	CXO/ACIS+	Jones,Forman,	$\sim$0.1–10.0 keV	8 h(X)	$\sim 2''$	NASA/CXC/SAO/
{Opt.	DSS	Murray				CJ/WF/SM/DSS
Soft X-ray	ROSAT/PSPC	Döbereiner, Böhringer, Biermann	0.2–2.0 keV	22.3 ks	$\geq 25''$	SD
Opt.	ESO 3.6m	Trinchieri	Hα + [N II]	2×600 s	$<2''$	GT

NGC 4472/M 49

X-ray	CXO/ACIS	Allen	0.6–8.0 keV	21.9 ks	$\sim 2''$	NASA/CXC/ Stanford U./SA+
Near-IR	2MASS		J,H,K$_S$	7.8 s	$2''$	2MASS

NGC 4676

Opt.	HST/ ACS/WFC	ACS ST	F475W,F606W, F814W	8.3 h		NASA/ ACS ST
Opt.	KPNO 2.1m	Hibbard	Hα+ [N II]	3×900 s	1.9$''$	JHi
Near-IR	UH88/ QUIRC	Hibbard	K	3×120 s	$\sim$0.9$''$	JHi
Radio	VLA (C+D)	Hibbard van Gorkom	21 cm line	21.5 h+6 h	$20'' \times 19''$	JHi

NGC 5194/M 51

X-ray	CXO/ ACIS	Kilgard	0.1–10 keV	12 h	$\sim 2''$	NASA/CXC/ Wesleyan U./RKi
Soft X-ray	ROSAT/HRI	Ehle, Beck	0.1–2.4 keV	16 ks		ME/RB
Mid-UV	HST/FOC		F275W	2277 s	$\sim$0.1$''$	DM
Opt.	HST/ACS/WFC		F435W, F555W, F658N, F814W	9 h		NASA/ESA/SB/HHT
Opt.	Pal. 1.5m/ WF-PFUEI		Hα	2500 s	$2''$	RR
Opt.	HST/WFPC2	Panagia				NP/NASA

Table 4 (*cont.*)

Galaxy/region	Tel./Instr.	Observ.	λ/ν/Energy/Filt.	Exp.	Res.	Image source
Opt.	HST/WFPC2		F439W,F555W, F814W,Hα			HHT/NASA
Near-IR	HST/ NICMOS	Scoville+	Pα, F1110W, F160W,F222M			NS/NASA
{Near-IR+ {Mid-IR	Spitzer/ IRAC	Kennicutt	3.6,4.5,5.6, 8.0 μm	4 m		NASA/JPL-Caltech/ RK+
Radio	OVRO	Rand	CO ($J = 1 \rightarrow 0$)		$9'' \times 7''$	RR
Radio	Eff. 100m+ VLA(D)	Neininger	6.2 cm/ 4.86 GHz	5+9 h	$12''$	NN
Radio	VLA (A)	Crane, v. d. Hulst	6 cm/ 4.86 GHz		$0.55'' \times 0.45''$	NRAO
Radio	VLA (B,D)	Rots	21 cm line		$34''$	AR

NGC 520

Galaxy/region	Tel./Instr.	Observ.	λ/ν/Energy/Filt.	Exp.	Res.	Image source
Opt.	KPNO 0.9m	Hibbard	Hα + [N II]	4×900 s	$2.8''$	JHi
Opt.	KPNO 0.9m	Hibbard	R	5×600 s	$2.8''$	JHi
Near-IR	KPNO 1.3m/ IRIM	Bushouse	2.2 μm	60 s	$2.7''$	AS
Radio	VLA (C+D)	Hibbard, van Gorkom	21 cm line	21.5 h + 7 h	$24'' \times 24''$	JHi

NGC 1275

Galaxy/region	Tel./Instr.	Observ.	λ/ν/Energy/Filt.	Exp.	Res.	Image source
X-ray	CXO/ACIS	Fabian	0.3–7.0 keV	54 h	$2''$	NASA/CXC/IoA/AF+
Opt.	HST/ ACS/WFC	Fabian+	F435W,F550M, F625W			NASA/ESA/HHT/ AF+
Radio	VLA		43,8 GHz			NRAO/AUI
Radio	VLA (A)	Pedlar	90 cm/0.3 GHz		$5.5'' \times 4.9''$	AP

NGC 1316

Galaxy/region	Tel./Instr.	Observ.	λ/ν/Energy/Filt.	Exp.	Res.	Image source
{Far-UV+ {Near-UV+ {Opt.	GALEX		1400–1700, 1800–2750 Å red		$4.2''$	NASA/JPL-Caltech/ CTIO
Opt.	CTIO CS	Mackie	B	11700 s	$\sim4''$	GHM
Opt.	CTIO CS	Mackie	Hα+[N II]	8100 s	$\sim4''$	GHM
Opt.	HST/ACS		F435W,F555W, F814W	3.5 h		NASA,ESA,HHT
Radio	VLA	Fomalont	1.51 GHz		$14''$	NRAO
Radio	VLA	Fomalont+	1.4 GHz		$14''$	NRAO/AUI

NGC 4038/9

Galaxy/region	Tel./Instr.	Observ.	λ/ν/Energy/Filt.	Exp.	Res.	Image source
X-ray	CXO/ACIS	Fabbiano	0.3–0.65,0.65–1.5, 1.5–6.0 keV	117 h	$\sim2''$	NASA/CXC/+ SAO/GF+
Opt.	HST/ACS	Whitmore	F435W,F550M, F658N,F814W	4.9 h		NASA,ESA+ BW

Table 4 (*cont.*)

Galaxy/region	Tel./Instr.	Observ.	λ/ν/Energy/Filt.	Exp.	Res.	Image source
Opt.	CTIO 0.9m	Hibbard	B,V,R	2×600 s	1.8″	JHi/PG
Near-IR	2MASS		J,H,K_S	7.8 s	2″	2MASS
{Near-IR+	Spitzer/	Wang+	3.6,4.5,5.8,	160 s		NASA/JPL-Caltech/
{Mid-IR	IRAC		8.0 μm			ZW+
Radio	VLA (C+D)	Hibbard	21 cm line	8 h+3.5 h	20″ × 19″	JHi

NGC 7252

Opt.	HST/WFPC	Whitmore	F555W, F785LP	2×700 s each	0.1″	NASA/BW+
Opt.	KPNO 2.1m	Hibbard	R	300 s	2″	JHi
Radio	VLA (C+D)	Hibbard, v. Gorkom	21 cm line	22.5+7 h	27″ × 16″	JHi

NGC 253

X-ray	XMM-Newton	Bauer	0.2–0.5,0.5–1.0, 1.0–2.0 keV		20″	MBa/MPE/ESA
X-ray	CXO/ACIS		0.2–1.5 keV	4 h	~2″	NASA/SAO/CXC
{Far-UV+	GALEX		1400–1700,		4.2″	NASA/JPL-Caltech
{Near-UV			1800–2750 Å			
Opt.	CTIO CS	Mackie	B	1800 s	~4″	GHM
Opt.	HST/ACS	Dalcanton	F450W,F555W, F814W			NASA/ESA/JD/BWi
Opt.	CTIO CS	Mackie	$H\alpha$+[N II]	1800 s	~4″	GHM
Near-IR	2MASS		J,H,K_S	7.8 s	2″	2MASS
Radio	OVRO	Canzian	CO ($J = 1 \rightarrow 0$)	16 h	5″ × 9″	BC
Radio	ATCA	Koribalski	21 cm line	3×12 h	30″	BK

NGC 3034/M 82

X-ray	CXO/ACIS	Fabbiano	0.1–10.0 keV	13 h	~2″	NASA/SAO/ GF+
Mid-UV	UIT		A1	270.5 s	3″	NDADS
Opt.	Subaru/ FOCAS		B,V,$H\alpha$	30,25, 120 s		Subaru
Near-IR	Stew. 2.3m/ NICMOS2	Rieke[2]	H		1″	KM/GR/ MRi/DK
{Near-IR+	Spitzer/	Engelbracht	3.6,4.5,	1 h		NASA/JPL-Caltech/
{Mid-IR	IRAC		5.8–8.0 μm			CE+
Radio	JCMT/ BSISR	Tilanus, Tacconi, Sutton	CO ($J = 3 \rightarrow 2$)		14″	RT
Radio	OVRO	Seaquist+	0.3 cm/ 91.97 GHz	33 h	3″ × 3.5″	ESe

Table 4 (*cont.*)

Galaxy/region	Tel./Instr.	Observ.	λ/ν/Energy/Filt.	Exp.	Res.	Image source
NGC 5236/M 83						
X-ray	XMM-Newton	Watson	0.2–0.8,0.8–1.5, 1.5–12 keV	57,74 ks	20″	RWi/PR/ESA
{Far-UV+	GALEX		1400–1700,		4.2″	NASA/JPL-Caltech/
{Near-UV+			1800–2750 Å			
{Radio	VLA		21 cm line			VLA/MPIA
Opt.	HST/WFC3	O'Connell	F336W,F555W, F814W,F502N, F657N	2.3 h		NASA/ESA/ RO/WSOC/ ESO
Opt.	AAT f/8/ TAURUS II	Ryder, Dopita, Sutherland	Hα	1000 s	2″	SDR/ MD/RS
Opt.	NOT	Larsen	I		1.5″	SL
Mid-IR	ISO/ISOCAM	Roussel+	7 μm		5.7″	HR
Radio	Eff. 100m+ VLA	Beck, Neininger	6 cm/4.9 GHz		10″	RB
Radio	ATCA		21 cm line		86″ × 61″	BK
NGC 1068/M 77						
{X-ray+	CXO/ACIS		0.4–0.8,	13 h(X)	~2″	NASA/CXC/MIT/
{			0.8–1.3 keV			UCSB/PO+
{Opt.	HST					NASA/STScI/AC+
Opt.						NASA/STScI
Opt.	Lowell 1.1m	Keel	Hα+[N II]	600 s	~2″	WK
Radio	BIMA	Regan	CO ($J = 1 \rightarrow 0$)		6″	MRe
Radio	VLA	Wilson, Ulvestad	6 cm/4.9 GHz			AW/JU
NGC 1365						
{Far-UV+	GALEX		1400–1700,		4.2″	NASA/JPL-Caltech/
{Near-UV			1800–2750 Å			SSC
Near-IR	HST/NICMOS	Carollo	1.6 μm			NASA/ESA/CMC
Radio	ATCA	Norris, Forbes	6 cm/ 4.8 GHz	4 h	3.1″ × 1.6″	RN
Radio	VLA	v. Moorsel, Jorsater	21 cm line	60 h	6.2″ × 5.0″	vM/SJ
NGC 3031/M 81						
{Far-UV+	GALEX	Huchra	1400–1700,		4.2″	NASA/JPL-Caltech/
{Near-UV			1800–2750 Å			JH
Opt.	HST/ACS		F435W,F606W, F814W			NASA/ESA/HHT
{Mid-IR+	Spitzer/	Gordon	24,70,	80,40,		NASA/JPL-Caltech/
{Far-IR	MIPS		160 μm	8 s		KG

Table 4 (*cont.*)

Galaxy/region	Tel./Instr.	Observ.	λ/ν/Energy/Filt.	Exp.	Res.	Image source
Radio	Eff. 100m	Krause	2.8 cm		75″	MK
Radio	VLA (D,C/D)	Condon	20 cm/ 1.49 GHz		~0.9′	NRAO
Radio	VLA (B,C,D)	Adler, Westpfahl	21 cm line	>60 h	12″	NRAO/AUI

NGC 4258/M 106

{X-ray+	CXO/ACIS			4 h (X)		X: NASA/CXC/
{						U. Maryland/AW+;
{Opt.+	DSS					O: Pal. Obs./DSS;
{IR+	Spitzer					IR: NASA/JPL-Caltech;
{Radio	VLA		1.4 GHz			R: NRAO/AUI/NSF
{Far-UV+	GALEX		1400–1700,		4.2″	NASA/JPL-Caltech
{Near-UV			1800–2750 Å			
Opt.	FLWO 1.2m	McLeod	B	120 s	2–3″	KM
Radio	BIMA	Regan+	CO ($J = 1 \rightarrow 0$)		6″	MRe

3C 273

X-ray	CXO/HRC	Marshall	0.1–10.0 keV		0.5″	NASA/CXC/SAO/HM+
Opt.	HST/WFPC2	Bahcall	F606W	30 m		NASA/JB

NGC 4486/M 87

X-ray	CXO/ACIS	Forman		146 h		NASA/CXC/CfA/WF
{X-ray+	CXO/	Marshall	0.1–10.0 keV	38 ks		NASA/CXC/MIT/
{	ACIS					HM+
{Opt.+	HST	Perlman				NASA/STScI/UMBC/
{						EP+
{Radio	VLA	Zhou, Owen,				FZ/FO/NRAO/
{		Biretta				JBi/STScI
Mid-UV	HST/FOC		F220W	20 m	~0.1″	DM
Mid-UV	HST/ACS		F220W,F250W			NASA/ESA/JM
Opt.	DSS		~V		~1.7″	DSS
Opt.	HST/WFPC2	Ford+	Hα			NASA/HF+
Near-IR	2MASS		J,H,K$_s$	7.8 s	2″	2MASS
Radio	VLA (D)	Owen+	90 cm/330 MHz		8″	NRAO/AUI/FO/ JE/NKa
Radio	VLBA	Junor	43 GHz		0.33 × 0.12 mas	WJ/JBi/ML/ NRAO/AUI

NGC 4594/M 104

{X-ray+	CXO/ACIS			4 h 33 m		X: NASA/UMass/QDW+,
{Opt.+	HST					O: NASA/STScI/
{						AURA/HHT,
{IR	Spitzer					IR: NASA/JPL-Caltech/
{						U. AZ/RK/SINGS Team

Table 4 (*cont.*)

Galaxy/region	Tel./Instr.	Observ.	λ/ν/Energy/Filt.	Exp.	Res.	Image source
Opt.	ESO VLT/ FORS1	Barthel	V,R, I	120,120, 240 s	0.7 ″	ESO/PB/MN
Near-IR	2MASS		J,H,K_s	7.8 s	2 ″	2MASS
{Near-IR+	Spitzer		3.6,4.5,5.8,	240 s/filt.		NASA/JPL-Caltech/
{Mid-IR			8.0 μm			RK/SINGS Team
Radio	VLA (D)	Krause	6.2 cm/4.8 GHz	16 h	23 ″	MK/RW

NGC 5128

Galaxy/region	Tel./Instr.	Observ.	λ/ν/Energy/Filt.	Exp.	Res.	Image source
{X-ray+	CXO/HRC	Karovska	0.1–10.0 keV	4.7 h	0.5 ″	NASA/CXC/MKa+
{Opt.+			∼V		∼1.7 ″	DSS/STScI
{Radio+	VLA	Schiminovich	21 cm line			NRAO/VLA/JvG/DS+
{Radio	VLA	Condon	1.43 GHz			NRAO/VLA/JC+
{Far-UV+	GALEX		1400–1700,		4.2 ″	NASA/JPL-Caltech/
{Near-UV			1800–2750 Å			SSC
Opt.	DSS		∼V		∼1.7 ″	DSS
{Opt.+	HST/	Schreier				ESc/NASA
{	WFPC2					
{Near-IR	HST/		Pα			
{	NICMOS					
Near-IR	2MASS		J,H,K_s	7.8 s	2 ″	2MASS
Near-IR+Mid-IR	Spitzer		5.8,8.0 μm	60 s/filt.		NASA/JPL-Caltech/JK
Radio	SHEVE (VLBI)	Tingay+	⌀3 cm/ 8.4 GHz	∼12 h	3.3 × 1.8 mas PA 26.2°	ST/SHEVE Team
Radio	SHEVE (VLBI)	SHEVE Team	3+6 cm/ 8.4+4.8 GHz		3.0 mas	ST/SHEVE Team
Radio	P64	Junkes+	6.3 cm		4.3 ′	NJ/RH
Radio	VLA (C)		1.4 GHz			NRAO/AUI/ JBu/ESc/EF

A1795 #1

Galaxy/region	Tel./Instr.	Observ.	λ/ν/Energy/Filt.	Exp.	Res.	Image source
X-ray	CXO/ACIS	Fabian	0.1–10 keV	10.8 h	∼2 ″	NASA/IoA/AF+
Opt.	DSS		∼V		∼1.7 ″	DSS
Opt.	HST/WFPC2		¶(V+R)	1780 s	∼0.1 ″	JP
Radio	VLA		3.6 cm		0.37 ″	NRAO/AUI/NSF/ FO/JPG

Arp 220

Galaxy/region	Tel./Instr.	Observ.	λ/ν/Energy/Filt.	Exp.	Res.	Image source
X-ray	CXO/ACIS	McDowell	0.1–10 keV	15.6 h	∼2 ″	NASA/SAO/ CXC/JMcD
Opt.	HST ACS/WFC		F435W, F814W	33 m		NASA/ESA/HHT/AE
Opt.	HST/WFPC	Shaya, Dowling	V,R, I	800,780, 900 s	∼0.1 ″	ES/DD/ WFPC
Near-IR	HST/ NICMOS	Thompson+				NASA RTh+
Radio	VLA (A)	Norris	6 cm/ 4.9 GHz	8 h	0.48 ″ × 0.32 ″	RN

Table 4 (*cont.*)

Galaxy/region	Tel./Instr.	Observ.	λ/ν/Energy/Filt.	Exp.	Res.	Image source
Cygnus A						
X-ray	CXO/ACIS	Wilson	0.1–10.4 keV	9 h	$\sim 2''$	NASA/Maryland/ AW+
Opt.	DSS		$\sim$V		$\sim 1.7''$	DSS
Near-IR	Keck		K′		$\sim 0.05''$	LLNL/WMK
Radio	VLA (A,B, mixed)	Perley	6 cm/ 5 GHz			NRAO
Radio	VLBI MkIII	Carilli	6 cm/5 GHz		2 mas	NRAO

NOTES TO TABLE 4

- **Observatories:** CTIO, Cerro Tololo Inter-American Observatory; ESO, European Southern Observatory; FLWO, Fred Lawrence Whipple Observatory; KPNO, Kitt Peak National Observatory; Lowell, Lowell Observatory; McDonald, McDonald Observatory; MDM, Michigan Dartmouth MIT; OVRO, Owens Valley Radio Observatory; Pal., Palomar Observatory; Stew., Steward Observatory.

- **Telescopes:** AAT, Anglo-Australian Telescope; ATCA, Australia Telescope Compact Array; BIMA, Berkeley-Illinois-Maryland Association millimeter interferometer; BS, Burrell Schmidt; CFHT, Canada France Hawaii Telescope; Columb. 1.2m, Columbia 1.2 m Millimeter-Wave Telescope; CS, U. Michigan Curtis Schmidt; C.W.R.U., Case Western Reserve University; CXO, Chandra X-ray Observatory; DSS, (Digitized Sky Survey – either UK Schmidt or Oschin Schmidt); Eff. 100 m, Effelsberg 100 m; GALEX, Galaxy Evolution Explorer; HST, Hubble Space Telescope; IRAM, Institut de Radioastronomie Millimetrique; IRAS, Infrared Astronomical Satellite; ISO, Infrared Space Observatory; JCMT, James Clerk Maxwell Telescope; Keck, W.M. Keck Observatory; NANTEN, NANTEN Submillimeter Observatory; NOT, Nordic Optical Telescope; P64, Parkes 64 m Radio Telescope; ROSAT, Röntgen Observatory Satellite; SHEVE, Southern Hemisphere VLBI Experiment; Spitzer, Spitzer Space Telescope; Subaru, 8.2 m Subaru Telescope; Swift, Swift Gamma-Ray Burst Mission; UH88, University of Hawaii 88 inch; UIT, Ultraviolet Imaging Telescope; var., various baselines (see references); VLA, Very Large Array; VLBA, Very Long Baseline Array; VLBI, Very Long Baseline Interferometer; VLT, Very Large Telescope; WIYN, WIYN Consortium (U. Wisconsin, Indiana University, Yale University, and the National Optical Astronomy Observatories NOAO) 3.5 m; WSRT, Westerbork Synthesis Radio Telescope; XMM-Newton, X-ray Multi-Mirror Mission-Newton Observatory; 2MASS, 1.3 m, at Mt. Hopkins, AZ, and at CTIO, Chile; 30, 30 inch telescope; ⊘The 8.4 GHz image of NGC 5128 was made from data obtained with 15 radio telescopes (8 telescopes of the NRAO's VLBA and 7 telescopes of the Southern Hemisphere VLBI Experiment array in Australia and South Africa).

- **Instrument:** ACIS, AXAF CCD Imaging Spectrometer; ACS/WFC, Advanced Camera for Surveys (Wide Field Camera); BSISR, Berkeley 345 GHz SIS Receiver; EPIC MOS, European Photon Imaging Cameras; FOC, Faint Object Camera; FOCAS, Faint Object Camera And Spectrograph; FORS1, FOcal Reducer/low dispersion Spectrograph; HRC, High Resolution Camera; HR Cam, High Resolution Camera; HRI, High Resolution Imager; IRAC, Infrared Array Camera; IRIM, Infrared Imager; ISOCAM, ISO Camera; ISOPHOT, ISO Photopolarimeter; MIPS, Multiband Imaging Photometer; NICMOS, Near Infrared Camera and Multi-Object Spectrometer array; NICMOS2, Near Infrared Camera and Multi-Object Spectrometer array #2; PFC, Prime Focus Camera; PSPC, Position Sensitive Proportional Counter; QUIRC, QUick InfRared Camera 1024 × 1024 pixel HgCdTe; RFP, Rutger Fabry–Perot; TAURUS II, Fabry–Perot Interferometer/Focal Reducer for Direct Imaging; UVOT, UV/Optical Telescope; WF-PFUEI, Wide Field – Prime Focus Universal Extragalactic Instrument; WFC3, Wide Field Camera 3; WFPC, Wide Field Planetary Camera; WFPC2, Wide Field Planetary Camera 2.

- **Observer:** ACS ST, Advanced Camera for Surveys Science Team; SHEVE, Southern Hemisphere VLBI Experiment; var., Various; +, *et al.*

- **Wavelength/Frequency/Energy/Filter:** A1, $\lambda_c \sim 2800$ Å; A5, $\lambda_c \sim 2500$ Å; B1, $\lambda_c \sim 1500$ Å; B5, $\lambda_c \sim 1600$ Å; B, $\lambda_c \sim 4400$ Å; CO ($J = 1 \rightarrow 0$), 2.6 mm line; CO ($J = 3 \rightarrow 2$), 0.88 mm line; filters of the form F✭✭✭X or F✭✭✭✭X are UV, optical and near-IR filters where ✭✭✭ or ✭✭✭✭ is the approximate central wavelength, λ_c in nm, and X is N, M, W or LP for narrow, medium, wide or long-pass bandwidths respectively; F160BW* – despite 12,000 s of integration the WFPC2 F160BW exposure of NGC 224/M 31 had little if any signal recorded; Hα, narrow band filter, 6563 Å line; Hα+[N II], narrow band filter, 6563 Å, 6548,6583 Å lines; I, $\lambda_c \sim 8200$ Å; J,H,K$_S$, 2MASS J 1.2 μm, H 1.6 μm, K$_S$ 2.1 μm; K,

$\lambda_c \sim 2.2$ μm; K′, $\lambda_c \sim 2.1$ μm; K$_S$, $\lambda_c \sim 2.17$ μm; Pα, Paschen α 1.87 μm; R, $\lambda_c \sim 6500$ Å; U, $\lambda_c \sim 3600$ Å; UVW1, $\lambda_c \sim 1928$ Å; UVM2, $\lambda_c \sim 2246$ Å; UVW2, $\lambda_c \sim 2600$ Å; V, $\lambda_c \sim 5500$ Å; ¶(V+R), the average of F555W and F702W images with an elliptical model subtracted.

- **Exposure:** av., average; d, days; filt., filter; h, hours; ks, kiloseconds; m, minutes; s, seconds; 22 × 500 s, a series of 22 500 s exposures; var., various times.

- **Resolution:** ′ arc minutes; ″ arc seconds; mas, milli-arcseconds.

  The angular resolution of each image is typically measured as full width half maximum (FWHM) which is the width of the point response function (PRF) at half the maximum intensity. Radio observations are usually described in terms of half power beam width (HPBW) which is measured in a similar manner to FWHM except that the radio beam shape can be asymmetrical in which case two widths may be quoted, along with a position angle. †Indicates that IRAS maximum correlation method (MCM; Aumann *et al.* 1990) images are shown. The iteration 20 MCM image at 60 μm has a FWHM Cross-scan × In-scan of $62'' \times 41''$.

- **Image Source:** AC, A. Capetti; ACS ST, Advanced Camera for Surveys Science Team; AE, A. Evans; AF, A. Fabian; AH, A. Hughes; AH/TW/JO/JP/EM/MAGMA, A. Hughes, T. Wong, J. Ott, J. Pineda, E. Muller and the MAGMA collaboration; AM, A. Marlowe; AP, A. Pedlar; AR, A. Rots; AS, A. Stanford; AUI, Associated Universities, Inc.; AURA, Association of Universities for Research in Astronomy Inc.; AW, A. Wilson; AW/JU, A. Wilson, J. Ulvestad; AZ, A. Zezas; BC, B. Canzian; BK, B. Koribalski; BS, B. Savage; BW, B. Whitmore; BWi, B. Williams; CE, C. Engelbrecht; CfA, (Harvard-Smithsonian) Center for Astrophysics; CH, C. Howk; CJ, C. Jones; CMC, C. M. Carollo; CN, Ch. Nieten; CS, C. Smith; CSIRO, Commonwealth Scientific and Industrial Research Organisation; CTIO, Cerro Tololo Inter-American Observatory; CXC, Chandra X-ray Center; DD, D. Dowling; DG, D. Garnett; DM, D. Maoz; DP, D. Pooley; DS, D. Schiminovich; DSS, Digitized Sky Survey; EB, E. Brinks; EF, E. Feigelson; EG, E. Grand; ES, E. Shaya; ESc, E. Schreier; ESe, E. Seaquist; ESA, European Space Agency; ESO, European Southern Observatory; EMB, E. M. Berkhuijsen; EP, E. Perlman; FB, F. Bresolin; FO, F. Owen; FZ, F. Zhou; GB, G. Bothun; GF, G. Fabbiano; GH, G. Helou; GHM, G. Mackie; GR, G. Risaliti; GT, G. Trinchieri; HF, H. Ford; HGM, G. Meurer; HH, H. Hippelein; HHT, Hubble Heritage Team; HM, H. Marshall; HR, H. Roussel; IoA, Institute of Astronomy; INAF, Instituto Nazionale Di Astrofisica; IPAC, Infrared Processing and Analysis Center; IRAM, Institut de Radioastronomie Millimetrique; JB, J. Bahcall; JBi, J. Biretta; JBu, J. Burns; JC, J. Condon; JD, J. Dalcanton; JE, J. Eilek; JH, J. Hutchings; JHi, J. Hibbard; JHi/PG, J. Hibbard, P. Guhathakurta; JHU, Johns Hopkins University; JI, J. Irwin; JK, J. Keene; JM, J. Madrid; JMo, J. Mould; JM-A, J. Maiz-Apellanix; JMcD, J. McDowell; JP, J. Pinkney; JPG, J. P. Ge; JPL, Jet Propulsion Laboratory; JPL-Caltech, Jet Propulsion Laboratory-California Institute of Technology; JT, J. Trauger; JvG, J. van Gorkom; JW, J. Westphal; KG, K. Gordon; KK, K. Kuntz; KM, K. McLeod; KM/GR/MRi/DK, K. McLeod, G. Rieke, M. Rieke, D. Kelly; LLNL, Lawrence Livermore National Laboratory; LSS, L. Staveley-Smith; Maryland, University of Maryland; MBa, M. Bauer; MCELS, Magellanic Cloud Emission Line Survey Project; ME, M. Ehle; ME/RB, M. Ehle, R. Beck; MG, M. Guelin; MH, M. Haas; MIT, Massachusetts Institute of Technology; MKa, M. Karovska; MK, M. Krause; MK/RW, M. Krause, R. Wielebinski; ML, M. Livio; MM, K. Michael Merrill; MPE, Max-Planck-Institut fur Extraterrestrische Physik; MPIA, Max Planck Institute for Astronomy; MPIfR, Max-Planck-Institut for Radio Astronomy; MRe, M. Regan; MR, M. Rubio; NASA, National Aeronautics and Space Agency; ND, N. Duric; NDADS, NASA Data Archive and Distribution Service; NJ/RH, N. Junkes, R. Haynes; NKa, N. Kassim; NK, N. Killeen; NN, N. Neininger; NOAO, National Optical Astronomy Observatory; NP, N. Panagia; NRAO, National Radio Astronomy Observatories (CD-ROM); NS, N. Scoville; NSF, National

Table 4 (*cont.*)

NOTES TO TABLE 4 (*cont.*)

Science Foundation; NW, N. Walborn; OCIW, Observatories of the Carnegie Institute of Washington; Pal. Obs., Palomar Observatory; PBa, P. Bamby; PB/MN, P. Barthel, M. Neeser; PG/PC/SR, P. Guhathakurta, P. Choi, S. Raychaudhury; PH, P. Hoernes; PK, P. Kahabka; PO, P. Ogle; PR, P. Rodriguez; QDW, Q.D. Wang; RB, R. Beck; RBa, R. Barba; RDiS, R. DiStefano; RK, R. Kennicutt; RKi, R. Kilgard; RN, R. Norris; RO, R. O'Connell; RR, R. Rand; RT, R. Tilanus; RTh, R. Thompson; RW, R. Walterbos; RWi, R. Willatt; SA, S. Allen; SAGE, SAGE Legacy Team; SAO, Smithsonian Astrophysical Observatory; SB, S. Beckwith, SC+, S. Carpano, J. Wilms, M. Schirmer, E. Kendziorra; SD, S. Döbereiner; SDR/MD/RS, S. Ryder, M. Dopita, R. Sutherland; SHEVE, Southern Hemisphere VLBI Experiment; SI, S. Immler; SINGS Team, Spitzer Infrared Nearby Galaxies Survey Team; SK, S. Kim; SL, S. Larsen; SM/GB, S. McGaugh, G. Bothun; SM, S. Murray; SS, S. Snowden; SSC, Spitzer Science Center; ST, S. Tingay; Stanford U., Stanford University; STr, S. Trudolyubov; STScI, Space Telescope

Science Institute; Subaru, Subaru Telescope, National Astronomical Observatory of Japan (NAOJ); Swift, Swift Observatory; TL, T. Lauer; UA, U. Alabama; UCSB, University of California Santa Barbara; UK/RH, U. Klein, R. Haynes; UM, University of Michigan; UMBC, University of Maryland Baltimore Counties; U. Mass, University of Massachusetts; Univ. of Ariz., University of Arizona; USRDA, U.S. ROSAT Data Archive; VLA, Very Large Array; vM/SJ, G. van Moorsel, S. Jorsater; Wesleyan U., Wesleyan University; WF, W. Forman; WFPC, WFPC Team; WFPC IDT, WFPC Investigation Definition Team; WIYN, WIYN Consortium (U. Wisconsin, Indiana University, Yale University, and the National Optical Astronomy Observatories); Wisc., U. Wisconsin; WJ, W. Junor; WK, W. Keel; WMK, W.M. Keck Observatory; WP, W. Pietsch; WSOC, WFC3 Science Oversight Committee; YCC, Y.-C. Chu, ZL, Z. Li; ZW, Z. Wang; 2MASS, 2MASS/UMass/IPAC-Caltech/NASA/NSF; +, *et al.*

APPENDIX D
Cross-reference list (Table 5)

Galaxy	Optical	X-ray	Infrared (IRAS)	Radio
NGC 224/M 31	"Andromeda" [NGC 206]		00400 + 4059	
NGC 253			00450 − 2533	PKS 0045 − 25
NGC 300			00523 − 3756 00525 − 3757	
NGC 520	Arp 157, VV 231 [UGC 957]		01219 + 0331	
NGC 598/M 33			01310 + 3024	
	[NGC 604]			
NGC 891				
NGC 1068/M 77			02401 − 0013	3C 71, PKS 0240 − 002
NGC 1275			03164 + 4119	Perseus A, 3C 84, 4C 41.07
NGC 1316			03208 − 3723	Fornax A, PKS 0320 − 374
NGC 1365			03317 − 3618	PKS 0331 − 363
NGC 1399				PKS 0336 − 355
NGC 2915			09265 − 7624	
NGC 3031/M 81			09514 + 6918	
NGC 3034/M 82	Arp 337		09517 + 6954	3C 231, 4C 69.12
NGC 4038/9	"The Antennae", VV 245		11593 − 1835	PKS 1159 − 185
NGC 4258/M 106	VV 448			
NGC 4406/M 86			12234 + 1315	
NGC 4472/M 49	Arp 134			
NGC 4486/M 87		X Vir X-1	12282 + 1240	Virgo A, 3C 274, 4C 12.45
NGC 4594/M 104	"The Sombrero"		12373 − 1121	
NGC 4676(a,b)	"The Mice", Arp 242, VV 224		12437 + 3059	
NGC 5128			13225 − 4245	Centaurus A, PKS 1322 − 427
NGC 5194/M 51	"The Whirlpool", VV 1 [NGC 5195]		13277 + 4727	4C 47.36.1?

Table 5 (*cont.*)

Galaxy	Optical	X-ray	Infrared (IRAS)	Radio
NGC 5236/M 83			13341 − 2936	PKS 1334-296
NGC 5457/M 101	"The Pinwheel", VV 344 [NGC 5447-55-61-62]		14013 + 5435	4C 54.30.1
NGC 6822	"Barnard's Galaxy", IC 4895 [IC 1308]		19421 − 1455	
NGC 7252	"Atoms For Peace", AM 2217-245		22179 − 2455	
A1795 #1				4C 26.42
Arp 220	IC 4553/4, VV 540		15327 + 2340	
Cygnus A			19577 + 4035	3C 405, 4C 40.40
LMC	A 0524-69 [NGC 1850,54,56,72,98] [NGC 1903,13,16,39,43,84,87,94] [NGC 2004,05,19,31,58,65] [NGC 2070, 30Dor., R136] [NGC 2100,07]	[LMC X-1, CAL 83] [CAL 87]		
Malin 2	F568-6			
SMC	A 0051-73 [NGC 220,42,65] [NGC 330,46,61,71,76] [NGC 411,16,19,58]			
3C 273			12265 + 0219	PKS 1226 + 02, 4C 02.32

- Galaxy regions: Regions of galaxies (e.g. NGC 206 in NGC 224/M 31) or dynamically associated systems (e.g. NGC 5195 near NGC 5194/M 51) are listed in square ([ ]) brackets. Major NGC objects in the LMC and SMC are listed.

- Optical catalogues: VV #, Vorontsov-Velyaminov (1977); AM #, Arp and Madore (1987); UGC #, Nilson (1973); Arp #, Arp (1966).

- Radio catalogues: PKS #, Parkes, Ekers (1969); 3C #, Third Cambridge Catalogue, Edge *et al.* (1959), Bennett (1962); 4C #, Fourth Cambridge Catalogue, Pilkington and Scott (1965).

APPENDIX E
Common abbreviations and acronyms (Table 6)

Abbreviation, Acronym	
AAO	Anglo-Australian Observatory
AAT	Anglo-Australian Telescope
ADF	Astrophysics Data Facility
AGN	Active galactic nuclei
A. J.	*Astronomical Journal*
Ann. Rev. Astron. Astrophys.	*Annual Review Astronomy and Astrophysics*
Ap. J.	*Astrophysical Journal*
Ap. J. Suppl.	*Astrophysical Journal Supplement Series*
Astron. Ap.	*Astronomy and Astrophysics*
Astron. Ap. Suppl.	*Astronomy and Astrophysics Supplement Series*
ATCA	Australia Telescope Compact Array
ATNF	Australia Telescope National Facility
AU	Astronomical unit
AUI	Associated Universities, Inc.
Aust. J. Physics	*Australian Journal of Physics*
AURA	Association of Universities for Research in Astronomy Inc.
Blazar	Low (BL Lacs) and high (HPQs, OVVs) luminosity beamed AGN
BL Lac	BL Lacerta-like object
BLR	Broad line region
CCW	Counter-clockwise
CGRO	Compton Gamma Ray Observatory
CMB	Cosmic Microwave Background
COMPTEL	Imaging Compton Telescope
CTIO	Cerro Tololo Inter-American Observatory
CW	Clockwise
CXC	Chandra X-ray Center
CXO	Chandra X-ray Observatory
EGRET	Energetic Gamma Ray Experiment Telescope
ESA	European Space Agency
ESO	European Southern Observatory
EUVE	Extreme Ultraviolet Explorer
FOC	Faint Object Camera
FR I, II	Fanaroff–Riley Class I, II
FWHM	Full width half maximum
GALEX	Galaxy Evolution Explorer
GRB	Gamma ray burst
GSFC	Goddard Space Flight Center
HPQ	Highly polarized quasar
HRI	High-Resolution Imager
HST	Hubble Space Telescope
IMBH	Intermediate mass black hole
IPAC	Infrared Processing and Analysis Center
IRAS	Infrared Astronomical Satellite
IC	Index Catalogue

Abbreviation, Acronym	
ICM	Intracluster medium
ISM	Interstellar medium
ISO	Infrared Space Observatory
K	Kelvin
KPNO	Kitt Peak National Observatory
LG	Local Group
LINER	Low ionization nuclear emission-line regions
LMC	Large Magellanic Cloud
LSB	Low surface brightness
M #	Messier (Catalogue) entry
M.N.R.A.S.	*Monthly Notices of the Royal Astronomical Society*
NASA	National Aeronautics and Space Administration
NGC	New General Catalogue
NLR	Narrow line region
NOAO	National Optical Astronomy Observatories
NRAO	National Radio Astronomy Observatories
OVV	Optically violent variables
PAHs	Polycyclic aromatic hydrocarbons
P.A.S.P.	*Publications of the Astronomical Society of the Pacific*
PSPC	Position Sensitive Proportional Counter
Publ. Astron. Soc. Aust.	*Publications Astronomical Society of Australia*
Publ. Astron. Soc. Japan	*Publications Astronomical Society of Japan*
ROSAT	Röntgen Observatory Satellite
RSA	Revised Shapley–Ames (Catalog)
SED	Spectral energy distribution
SFR	Star-formation rate
SMBH	Supermassive black hole
SMC	Small Magellanic Cloud
SNe	Supernovae
SNR	Supernova remnant
solar	$\odot$
solar mass	$M_\odot$
SST	Spitzer Space Telescope
STScI	Space Telescope Science Institute
Sy 1, 2	Seyfert 1, 2
UIT	Ultraviolet Imaging Telescope
ULX	Ultraluminous X-ray (source)
USRDA	U.S. ROSAT Data Archive
UV	Ultraviolet
VLA	Very Large Array
VLBA	Very Long Baseline Array
VLBI	Very long baseline interferometry
WFPC2	Wide Field Planetary Camera 2
WSRT	Westerbork Synthesis Radio Telescope

APPENDIX F
Astronomical constants and units (Table 7)

Symbol	Meaning	Value
Å	ångström	10^{-10} m
AU	astronomical unit	1.496×10^{8} km
c	speed of light	2.998×10^{5} km s^{-1}
eV	electron volt	1.602×10^{-12} erg
G	gauss	1 Mx cm^{-2}
k	Boltzmann's constant	1.381×10^{-16} erg K^{-1}
$L_\odot$	solar luminosity (bolometric)	3.827×10^{33} erg s^{-1}
$M_\odot$	solar mass	1.989×10^{33} g
pc	parsec	3.086×10^{18} cm
pc	parsec	3.262 light-years
$''$	arcsecond	1/3600 of 1°
$'$	arcminute	1/60 of 1°

- Prefixes: p − pico, 10^{-12}; n − nano, 10^{-9}; μ − 10^{-6}; m(illi) − 10^{-3}; c(enti) − 10^{-2}; k(ilo) − 10^{3}; M(ega) − 10^{6}; G(iga) − 10^{9}; T(era) − 10^{12}.
- Bolometric − includes radiation over all wavelengths.

APPENDIX G
Other atlases of galaxies

Arp, H.C. 1966, Ap. J. Suppl., **14**, 1., also *Atlas of Peculiar Galaxies*, (Calif. Inst. Techn. Pasadena)
http://nedwww.ipac.caltech.edu/level5/Arp/frames.html

Elmegreen, D.M. 1981, Ap. J. Suppl., **47**, 229. (A Near-Infrared Atlas of Spiral Galaxies)
http://nedwww.ipac.caltech.edu/level5/Elmegreen/frames.html

Hartmann, D. and Burton, W.B. 1997, *Atlas of Galactic Neutral Hydrogen* (Cambridge University Press, Cambridge)

Hodge, P.W. 1981, *Atlas of the Andromeda Galaxy* (University of Washington Press, Seattle and London)
http://nedwww.ipac.caltech.edu/level5/ANDROMEDA_Atlas/frames.html

Jarrett, T.H. *et al.* 2003, A. J., **125**, 525. (2MASS Large Galaxy Atlas)
http://nedwww.ipac.caltech.edu/level5/Sept02/Jarrett/index.html

Laustsen, S., Madsen, C. and West, R.M. 1987, *Exploring the Southern Sky* (Springer-Verlag, Berlin)

Marcum, P.M. *et al.* 2001, Ap. J. Suppl., **132**, 129. (An Ultraviolet/Optical Atlas of Bright Galaxies)
http://nedwww.ipac.caltech.edu/level5/March01/Marcum/frames.html

Sandage, A. 1961, *Hubble Atlas of Galaxies* (Carnegie Institution of Washington, Washington, D.C.)

Sandage, A. and Tammann, G.A. 1981, *A Revised Shapley-Ames Catalog of Bright Galaxies* (Carnegie Institute of Washington, Washington, D.C.) (RSA)
http://nedwww.ipac.caltech.edu/level5/Shapley_Ames/frames.html

Sandage, A. and Bedke, J. 1988, Atlas of Galaxies Useful for Measuring the Cosmological Distance Scale, NASA SP-496, Washington, DC.

Scoville, N.Z. *et al.* 2000, A. J., **119**, 991. (NICMOS Imaging of Infrared-Luminous Galaxies)
http://nedwww.ipac.caltech.edu/level5/Scoville/frames.html

Vorontsov-Velyaminov, B.A. 1959, *The Atlas and Catalogue of Interacting Galaxies* (Moscow).
http://nedwww.ipac.caltech.edu/level5/VV_Cat/frames.html

APPENDIX H
Spectral energy distributions

In this appendix several spectral energy distributions (SEDs) comprising individual multiwavelength measurements of atlas galaxies are shown to illustrate differences in multiwavelength properties. The SEDs are from the NASA/IPAC Extragalactic Database (NED).

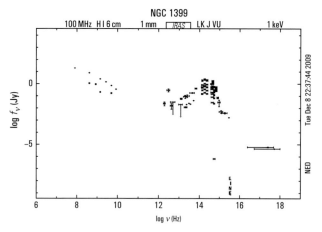

Figure H.1 SED of the normal galaxy NGC 1399. Wavelength (log scale in frequency) is plotted horizontally, and flux vertically (log scale f_ν in Jy). Credit: NED.

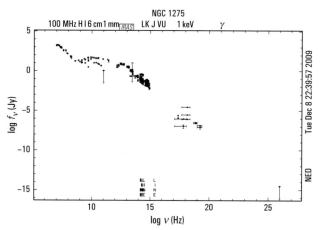

Figure H.2 SED of the merging galaxy NGC 1275. Credit: NED.

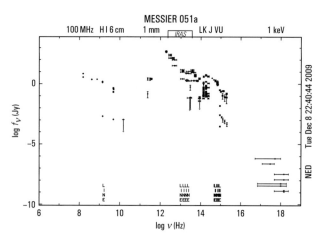

Figure H.3 SED of the interacting galaxy NGC 5194/M 51. Credit: NED.

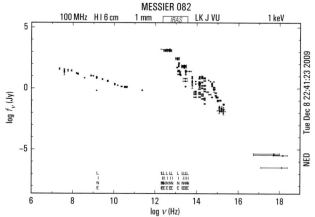

Figure H.4 SED of the starburst galaxy NGC 3034/M 82. Credit: NED.
A spectrum (with model fits) of the radio, submillimeter and far-IR regions of NGC 3034/M 82 is also shown in Figure 2.13.

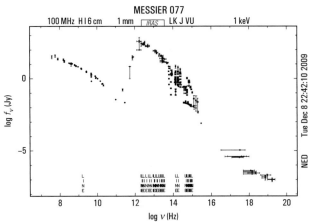

Figure H.5 SED of the Seyfert galaxy NGC 1068/M 77. Credit: NED.

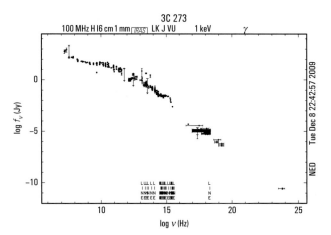

Figure H.6 SED of the quasar 3C 273. Credit: NED.

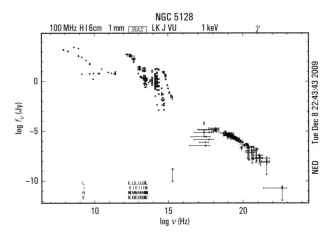

Figure H.7 SED of the radio galaxy NGC 5128 (Centaurus A). Credit: NED.

General text references

IN CHAPTERS 1 THROUGH 3

Abbe, C. 1867, M.N.R.A.S., **27**, 257.

Abell, G.O., Corwin, H.G. and Olowin, R.P. 1989, Ap. J. Suppl., **70**, 1.

Aldering, G.S. and Bothun, G.D. 1991, P.A.S.P., **103**, 1296.

Antonucci, R.R.J. and Miller, J.S. 1985, Ap. J., **297**, 621.

Antonucci, R.R.J. 1993, Ann. Rev. Astron. Ap., **31**, 473.

Arp, H.C. 1966, Ap. J. Suppl., **14**, 1. (also *Atlas of Peculiar Galaxies*, Calif. Inst. Techn. Pasadena)

Arp, H.C. and Madore, B.F. 1987, *A Catalogue of Southern Peculiar Galaxies and Associations* (Cambridge University Press, Cambridge), vols. 1 and 2.

Aumann, H.H., Fowler, J.W., and Melnyk, M. 1990, A. J., **99**, 1674.

Babcock, H.W. 1939, Lick Obs. Bull., 19, 41.

Beck, R. *et al.* 1996, Ann. Rev. Astron. Ap., **34**, 155.

Beckwith, S.V.W. *et al.* 2006, A. J., **132**, 1729.

Bell, E.F. 2003, Ap. J., **586**, 794.

Bennett, A. 1962, Mem. R. Astron. Soc., **68**, 163.

Blandford, R. and Rees, M. 1974, M.N.R.A.S., **169**, 395.

Bridle, A.H. *et al.* 1994, A. J., **108**, 766.

Bronfman, L. *et al.* 1988, Ap. J., **324**, 248.

Burton, M.G. *et al.* 2000, Ap. J., **542**, 359.

Calzetti, D. *et al.* 2007, Ap. J., **666**, 870.

Carilli, C.L. *et al.* 1991, Ap. J., **383**, 554.

Condon, J.J. 1992, Ann. Rev. Astron. Ap., **30**, 575.

Cox, P. *et al.* 1998, Ap. J., **495**, L23.

Davidson, K. and Humphreys, R.M. 1997, Ann. Rev. Astron. Ap., **35**, 1.

de Vaucouleurs, G. 1970, IAU Symposium 38, *The Spiral Structure of our Galaxy*, eds. Becker, W. and Contopolous, G. (Reidel, Dordrecht), p. 18.

Djorgovski, S.G. *et al.* 1997, Nature, **387**, 876.

Dopita, M.A. 1997, Publ. Astron. Soc. Aust., **14**, 230.

Dopita, M.A. 2005, in *The Spectral Energy Distribution of Gas-Rich Galaxies: Confronting Models with Data*, Proceedings of the International Workshop in Heidelberg, eds. Popescu, C.C. and Tuffs, R.J. AIP Conf. Ser.

Dreyer, J.L.E. 1888, New General Catalogue of Nebulae and Clusters, Mem. R. Astron. Soc. **49**, pt. 1.

Dreyer, J.L.E. 1895, Index Catalogue, Mem. R. Astron. Soc. **51**, 185.

Dreyer, J.L.E. 1908, Index Catalogue, Mem. R. Astron. Soc. **59**, 105.

Duncan, J.C. 1922, P. A. S. P., **34**, 290.

Dunkley, J. *et al.* 2009, Ap. J. Suppl., **180**, 306.

Dwek, E. and Barker, M.K. 2002, Ap. J. **575**, 7.

Edge, D.O. *et al.* 1959, Mem. R. Astron. Soc. **48**, 37.

Ekers, J.A. 1969, Aust. J. Phys. Suppl., **7**, 1.

Evans, D.S. 1870, Nature, **3**, 167.

Fanaroff, B.L. and Riley, J.M. 1974, M.N.R.A.S., **167**, 31.

Fender, R.P. *et al.* 1999, M.N.R.A.S., **304**, 865.

Fishman, G.J. 1995, P.A.S.P., **107**, 1145.

Ford, H.C. 1998, S.P.I.E., **3356**, 234.

Freeman, K.C. 1970, Ap. J., **160**, 811.

Fruscione, A. 1996, in *Astrophysics in the Extreme Ultraviolet*, eds. Bowyer, S. and Malina, R.F. (Kluwer, Dordrecht), p. 69.

Gallagher, J. and Fabbiano, G. 1990, in *Windows on Galaxies*, eds. Fabbiano, G., Gallagher, J.S., and Renzini, A., (Kluwer, Dordrecht), p. 1.

Gillessen, S. *et al.* 2009, Ap. J., **707**, L114.

Genzel, R., Eckart, A., Ott, T. and Eisenhauer, F. 1997, M.N.R.A.S., **291**, 219.

Heckman, T.M. 1980, Astron. Ap., **87**, 152.

Herschel. J.F.W. 1864, Philosophical Transactions of the Royal Society of London, **154**, 1.

Ho, L.C. 2008, Ann. Rev. Astron. Ap., **46**, 475.

Holtzman, J.A. *et al.* 1995, P.A.S.P., **107**, 156.

Hubble, E.P. 1925, The Observatory, **48**, 139.

Hughes, D.H. *et al.* 1998, Nature, **394**, 241.

Impey, C. and Bothun, G. 1997, Ann. Rev. Astron. Ap., **35**, 267.

Iye, M. *et al.* 2006, Nature, **443**, 186.

Jacoby, G.H., Hunter, D.A. and Christian, C.A., 1984, Ap. J. Suppl., **56**, 257.

Kant, I. 1755, *The Universal Natural History and Theory of Heavens.*

Kennicutt, R.C. 1983, Ap. J., **272**, 54.

Kennicutt, R.C. 1993, in *The Environment and Evolution of Galaxies*, eds. Shull, J.M. and Thronson, H.A. (Kluwer, Dordrecht), p. 533.

Kerr, F.J. Hindman, J.V. and Carpenter, M.S. 1957, Nature, **180**, 677.

Kessler, M.F. *et al.* 1996, Astron. Ap., **315**, L27.

Kraemer, K.E., and Jackson, J.M. 1999, Ap. J. Suppl., **124**, 439.

Krause, M. 2003, Proceedings of Sino-Germany Radio Astronomy Conference Radio Studies of Galactic Objects, Galaxies and AGNs. eds. J.L. Han, X.H. Sun, J. Yang, and R. Wielebinski. In Suppl. Issue, Acta Astronomica Sinica, **44**, 123.

Krennrich, F. *et al.* 1997, Ap. J., **481**, 758.

Leavitt, H.S. 1908, Annals of Harvard College Observatory, **60**, 87.

Lemke, D. *et al.* 1996, Astron. Ap., **315**, L64.

Longair, M.S. 2011, *High Energy Astrophysics*, (Cambridge University Press, Cambridge).

Marshall, H.L., Fruscione, A., and Carone, T.E. 1995, Ap. J., **439**, 90.

McConnachie, A.W. *et al.* 2009, Nature, **461**, 66.

Messier, C. 1784, Connaissance des Temps pour 1784, Paris 227.

Metzger, M.R. *et al.* 1997, Nature, **387**, 878.

Meurer, G.R. *et al.* 2009, Ap. J., **695**, 795.

Mezger, P.G., Duschl, W.J. and Zylka, R. 1996, Astron. Ap. Review, **7**, 289.

Miller, G.E. and Scalo, J.M. 1979, Ap. J. Suppl., **41**, 513.

Miller, J.M. 2007, Ann. Rev. Astron. Ap., **45**, 441.

Mirabel, I.F. and Rodriguez, L.F. 1994, Nature, **371**, 46.

Moffet, A.T. *et al.* 1972, in *External Galaxies and Quasi-Stellar Objects*, IAU Symp. 44, ed. Evans, D.S., (Reidel, Dordrecht), p. 228.

Murai, T. and Fujimoto, M. 1986, Astrophys. Space Sci., **119**, 169.

Nayfeh, M.H., Habbal, S.R. and Rao, S. 2005, Ap. J., **621** L121.

Neugebauer, G. *et al.* 1984, Ap. J., **278**, L1.

Nilson, P. 1973, Uppsala General Catalogue of Galaxies, Uppsala Obs. Ann. **6**.

Novaco, J.C. and Brown, L.W. 1978, Ap. J., **221**, 114.

Omont, A. 2007, Rep. Prog. Phys., **70**, 1099.

Oort, J.H. 1932, B.A.N., **6**, 249.

Öpik, E. 1922, Ap. J., **55**, 406.

Paczyński, B. 1995, P.A.S.P., **107**, 1167.

Panagia, N. *et al.* 1991, Ap. J., **380**, L23.

Paresce, F. 1990, *Faint Object Camera Instrument Handbook* (STScI, Baltimore).

Parsons, W. (The Earl of Rosse) 1850, Philosophical Transactions of the Royal Society, **140**, 499.

Perley, R.A., Willis, A.G. and Scott, J.S. 1979, Nature, **281**, 437.

Pilkington, J.D.H. and Scott, P.F. 1965, Mem. R. Astron. Soc., **69**, 183.

Porter, T.A. *et al.* 2009, astro-ph-arXiv:0907.0293

Primack, J.R. 2009, astro-ph-arXiv:0909.2021

Putman, M.E. *et al.* 1998, Nature, **394**, 752.

Putman, M.E. *et al.* 2003, Ap. J., **586**, 170.

Reynolds, R.J. *et al.* 1995 Ap. J., **448**, 715.

Risaliti, G. and Elvis, M. 2004, in *Supermassive Black Holes in the Distant Universe*, A.J. Barger ed., Astrophysics and Space Science Library Volume 308 (Kluwer, Dordrecht), p.187.

Rowan-Robinson, M. 1985, *The Cosmological Distance Ladder* (Freeman, New York), p. 153.

Salpeter, E.E. 1955, Ap. J., **121**, 161.

Sandage, A. 1961, *Hubble Atlas of Galaxies* (Carnegie Institution of Washington, Washington, D.C.)

Sandage, A. 1975, in *Galaxies and the Universe*, Sandage, A., Sandage, M., and Kristian, J., eds., (University of Chicago Press), Chapter 1.

Sandage, A. and Tammann, G.A. 1981, *A Revised Shapley–Ames Catalog of Bright Galaxies* (Carnegie Institute of Washington, Washington, D.C.) (RSA)

Scalo, J.M. 1986, Fundamentals of Cosmic Physics, **11**, 1.

Schmidt, M. 1963, Nature, **197**, 1040.

Schödel, R. *et al.* 2002, Nature, **419**, 694.

Seyfert, C.K. 1943, Ap. J., **97**, 28.

Slipher, V.M. 1917, The Observatory, **40**, 304.

Smith, S. 1936, Ap. J., **83**, 499.

Sneden, C. *et al.* 2003, Ap. J., **591**, 936.

Stecher, T.P. *et al.* 1992, Ap. J., **395**, L1.

Tanaka, Y. *et al.* 1995, Nature, **375**, 659.

Tilanus, R.P.J., van der Werf, P.P. and Israel, F.P. 2000, P.A.S.P., **217**, 177.

Trümper, J. 1983, Adv. Space Res., **2(4)**, 241.

Tully, R.B. 1988, *Nearby Galaxies Catalog* (Cambridge University Press, Cambridge).

Urry, C.M. and Padovani, P. 1995, P.A.S.P., **107**, 803.

van den Bergh, S. 1998, *Galaxy Morphology and Classification* (Cambridge University Press, Cambridge).

van der Kruit, P.C. 1973, Astron. Ap., **29**, 263.

Vijh, U.P., Witt, A.N. and Gordon, K.D. 2004, Ap. J., **606**, L65.

Vorontsov-Velyaminov, B.A. 1977, Astron. Ap. Suppl., **28**, 1.

Wannier, P. and Wrixon, G.T. 1972, Ap. J., **173**, L119.

Weinberg, M.D. and Blitz, L. 2006, Ap. J., **641**, L33.

Werner, M.W. *et al.* 2004, Ap. J. Suppl., **154**, 1.

Westphal, J.A. 1982, in *Space Telescope Observatory*, IAU Gen. Assembly 18, Patras, p. 28.

Wright, T. 1750, An original theory or new hypothesis of the universe, 1750: Facsimile reprint together with the first publication of *A Theory of the Universe* 1734, [by] Thomas Wright of Durham; introduction and transcription by M.A. Hoskin. (Reprint, 1971. Dawsons, London)

Wyse, R.F.G 2009, in *The Ages of Stars*, IAU Symp. 258, Mamajuk, E.E, Soderblom, D.R., Wyse, R.F.G. ed., (Cambridge University Press, Cambridge), p. 11.

Zwicky, F. 1937, Ap. J., **86**, 217.

Major image sources and references

Gamma ray
Fermi Gamma-ray Space Telescope

X-ray
Chandra X-ray Observatory Center
ROSAT U.S. Data Archive
XMM-Newton

Ultraviolet
Galaxy Evolution Explorer (GALEX)
Swift Telescope
Ultraviolet Imaging Telescope (UIT)

Optical
Digitized Sky Survey, STScI
MAST HST Data Archive

Infrared
Rice, W. *et al.* 1988, Ap. J. Suppl., **68**, 91.
Rice, W. 1993, A. J., **105**, 67.
Infrared Space Observatory (ISO)
IPAC
Spitzer Space Telescope
Two Micron All Sky Survey

Submillimeter
Columbia Southern Millmeter-Wave Telescope
James Clerk Maxwell Telescope (JCMT)

Radio
ATNF: Australia Telescope Compact Array (ATCA); Parkes Radio Telescope; Mopra
Condon, J.J. 1987, Ap. J. Suppl., **65**, 485.
Effelsberg Radio Telescope
National Radio Astronomy Observatory, and NRAO 1992, Images from the Radio Universe, CD-ROM.
Owens Valley Radio Observatory (OVRO)
Very Large Array (VLA)
Westerbork Radio Synthesis Telescope (WRST)

Selected references by galaxy

IN CHAPTER 4

NGC 224/M 31

Multiwavelength
Hodge, P. 1992, *The Andromeda Galaxy* (Kluwer, Dordrecht).

X-ray
Kong, A.K.H. *et al.* 2002, Ap. J., **577**, 738.
Supper, R. *et al.* 1997, Astron. Ap., **317**, 328.
Supper, R. *et al.* 2001, Astron. Ap., **373**, 63.

Ultraviolet
King, I.R., *et al.* 1992, Ap. J., **397**, L35.
King, I.R., Stanford, S.A. and Crane, P. 1995, A. J., **109**, 164.
Maoz, D. *et al.* 1996, Ap. J. Suppl., **107**, 215.
O'Connell, R.W. *et al.* 1992, Ap. J., **395**, L45.

Optical
Bender, R. *et al.* 2005, Ap. J., **631**, 280.
Ibata, R. *et al.* 1993, Nature, **412**, 49.
Lauer, T. *et al.* 1993, A. J., **106**, 1436.
Pellet, A. *et al.* 1978, Astron. Ap. Suppl., **31**, 439.

Infrared
Haas, M. *et al.* 1998, Astron. Ap., **338**, L33.
Rice, W. 1993, A. J., **105**, 67.

Radio
Beck, R., Berkhuijsen, E.M. and Hoernes, P. 1998, Astron. Ap. Suppl., **129**, 329.
Berkhuijsen, E.M., Beck, R. and Hoernes, P. 2003, Astron. Ap., **398**, 937.
Brinks, E. and Shane, W.W. 1984, Astron. Ap. Suppl., **55**, 179.
Condon, J.J. 1987, Ap. J. Suppl., **65**, 485.
Nieten, C. *et al.* 2006, Astron. Ap., **453**, 459.
Walterbos, R.A.M., Brinks, E. and Shane, W.W. 1985, Astron. Ap. Suppl., **61**, 451.

SMC

X-ray
Kahabka P. and Pietsch W., 1996, Astron. Ap., **312**, 919.
Kahabka P. *et al.* 1999, Astron. Ap. Suppl., **136**, 81.

Ultraviolet
Cheng, K.-P. *et al.* 1992, Ap. J., **395**, L29.

Optical
Kennicutt, R.C. *et al.* 1995, A. J., **109**, 594.
Sabbi, E. *et al.* 2009, Ap. J., **703**, 721.

Infrared
Rice, W. *et al.* 1988, Ap. J. Suppl., **68**, 91.

Radio
Filipovic, M.D. *et al.* 2002, M.N.R.A.S., **335**, 1085.
Haynes, R.F. *et al.* 1986, Astron. Ap., **159**, 22.
Haynes, R.F. *et al.* 1991, Astron. Ap., **252**, 475.
Rubio, M. *et al.* 1991, Ap. J., **368**, 173.
Stanimirovic, S. *et al.* 1999, M.N.R.A.S., **302**, 417.
Staveley-Smith, L. *et al.* 1995, Publ. Astron. Soc. Aust., **12**, 13.
Staveley-Smith, L. *et al.* 1997, M.N.R.A.S., **289**, 225.

NGC 300

X-ray
Carpano, S. *et al.* 2007, Astron. Ap., **466**, 17.
Read, A.M. and Pietsch, W. 2001, Astron. Ap., **373**, 473.

Ultraviolet
Maoz, D. *et al.* 1996, Ap. J. Suppl., **107**, 215.

Infrared
Davidge, T.J. 1998, Ap. J., **497**, 650.
Rice, W. 1993, A. J., **105**, 67.

Radio
Condon, J.J. 1987, Ap. J. Suppl., **65**, 485.
Puche, D., Carignan, C., and Bosma, A. 1990, A. J., **100**, 1468.

NGC 598/M 33

Multiwavelength
Bicay, M.D., Helou, G. and Condon, J.J. 1989, Ap. J., **338**, L53.
Rice, W. *et al.* 1990, Ap. J., **358**, 418.

X-ray
Long, K.S. *et al.* 1996, Ap. J., **466**, 750.
Orosz, J.A. 2007, Nature, **449**, 872.

Optical
Courtes, G. *et al.* 1987, Astron. Ap., **174**, 28.

Infrared
Rice, W. 1993, A. J., **105**, 67.

Radio
Condon, J.J. 1987, Ap. J. Suppl., **65**, 485.
Newton, K. 1980, M.N.R.A.S., **190**, 689.
Rosolowsky, E. 2007, Ap. J., **661**, 830.
Viallefond, F. *et al.* 1986, Astron. Ap. Suppl., **64**, 237.
Wilson, C.D. and Scoville, N. 1990, Ap. J., **363**, 435.

NGC 891

Optical
Howk, J.C. and Savage, B.D. 2000, A. J., **119**, 644.
Rand, R.J., Kulkarni, S.R. and Hester, J.J. 1990, Ap. J., **352**, L1.
Sciama, D.W. 1993, *Modern Cosmology and the Dark Matter Problem* (Cambridge University Press, Cambridge).

Radio
Condon, J.J. 1987, Ap. J. Suppl., **65**, 485.
Oosterloo, T, Fraternali, F. and Sancisi, R. 2007, A. J., **134**, 1019.

NGC 1399

X-ray

Rangarajan, F.V.N. *et al.* 1995, M.N.R.A.S., **272**, 665.
Irwin, J.A. *et al.* 2010, Ap. J., 712, L1.
Jones, C. *et al.* 1997, Ap. J., **482**, 143.

Ultraviolet

O'Connell, R.W. *et al.* 1992, Ap. J., **395**, L45.

Optical

Goudfrooij, P. *et al.* 1994a, Astron. Ap. Suppl., **104**, 179.
Goudfrooij, P. *et al.* 1994b, Astron. Ap. Suppl., **105**, 341.
Grillmair, C.J. *et al.* 1999, A. J., **117**, 167.
Killeen, N.E.B. and Bicknell, G.V. 1988, Ap. J., **325**, 165.
Lauer, T.R. *et al.* 1995, A.J., **110**, 2622.

Radio

Killeen, N.E.B., Bicknell, G.V. and Ekers, R.D. 1988, Ap. J., **325**, 180.

LMC

Gamma ray

Porter, T.A. *et al.* 2009, astro-ph-arXiv:0907.0293

X-ray

Snowden, S.L. and Petre, R. 1994, Ap. J., **436**, L123.
Trümper, J. *et al.* 1991, Nature, **349**, 579.

Ultraviolet

Cheng, K.-P. *et al.* 1992, Ap. J., **395**, L29.

Optical

Harris, J. and Zaritsky, D. 2009, A. J., **138**, 1243.
Kennicutt, R.C. *et al.* 1995, A. J., **109**, 594.

Infrared

Rice, W. *et al.* 1988, Ap. J. Suppl., **68**, 91.

Radio

Cohen, R.S. *et al.* 1988, Ap. J., **331**, L95.
Fukui, Y. *et al.* 2008, Ap. J. Suppl., **178**, 56.
Haynes, R.F. *et al.* 1986, Astron. Ap., **159**, 22.
Haynes, R.F. *et al.* 1991, Astron. Ap., **252**, 475.
Kim, S. *et al.* 1998, Ap. J., **503**, 674.
Klein, U. *et al.* 1989, Astron. Ap., **211**, 280.
Luks, T. and Rohlfs, K. 1992, Astron. Ap., **263**, 41.
Ott, J. *et al.* 2008, Publ. Astron. Soc. Aust., **25**, 129.

NGC 2915

Optical

Marlowe, A.T. *et al.* 1995, Ap. J., **438**, 563.
Marlowe, A.T. *et al.* 1997, Ap. J. Suppl., **112**, 285.
Meurer, G.R., Mackie, G. and Carignan, C. 1994, A. J., **107**, 2021.

Radio

Meurer, G.R. *et al.* 1996, A. J., **111**, 1551.

MALIN 2

Optical

Bothun, G.D. *et al.* 1990, Ap. J., **360**, 427.
McGaugh, S.S., and Bothun, G.D. 1994, A. J., **107**, 530.
McGaugh, S.S., Schombert, J.M., and Bothun, G.D. 1995, A. J., **109**, 2019.

Radio

Matthews, L.D., van Driel, W. and Monnier-Ragaigne, D. 2001, Astron. Ap., **365**, 1.

NGC 5457/M 101

X-ray

Jenkins, L.P. *et al.* 2005, M.N.R.A.S., **357**, 401.
Pence, W.D. *et al.* 2001, Ap. J., **561**, 189.
Snowden, S.L. and Pietsch, W. 1995, Ap. J., **452**, 627.

Optical

Scowen, P.A., Dufour, R.J. and Hester, J.J. 1992, A. J., **104**, 92.

Infrared

Devereux, N.A. and Scowen, P.A. 1994, A. J., **108**, 1244.
Hippelein, H. *et al.* 1996a, Astron. Ap., **315**, L79.
Hippelein, H. *et al.* 1996b, Astron. Ap., **315**, L82.
Rice, W. 1993, A. J., **105**, 67.

Radio

Condon, J.J. 1987, Ap. J. Suppl., **65**, 485.
Kamphuis, J., Sancisi, R. and van der Hulst, T. 1991, Astron. Ap., **244**, L29.
Kenney, J.D.P., Scoville, N.Z. and Wilson, C.D. 1991, Ap. J., **366**, 432.
Smith, D.A. *et al.* 2000, Ap. J., **538**, 608.

NGC 6822

Optical

Hodge, P. 1977, Ap. J. Suppl., **33**, 69.
Hodge, P., Lee, M.G. and Kennicutt, R.C. 1989, P.A.S.P., **101**, 32.
Wyder, T.K. 2001, A. J., **122**, 2490.

Infrared

Cannon, J.M. *et al.* 2006, Ap. J., **652**, 1170.
Gallagher, J.S. *et al.* 1991, Ap. J., **371**, 142.
Rice, W. 1993, A. J., **105**, 67.

Radio

de Blok, W.J.G. and Walter, F. 2000, Ap. J., **537**, L95.
Condon, J.J. 1987, Ap. J. Suppl., **65**, 485.
Gottesman, S.T. and Weliachew, L. 1977, Astron. Ap., **61**, 523.
Wilson, C.D. 1992, Ap. J., **391**, 144.

NGC 4406/M 86

X-ray

Forman, W. *et al.* 1979, Ap. J., **234**, L27.

Optical

Nulsen, P.E.J. and Carter, D. 1987, M.N.R.A.S., **225**, 939.

Trinchieri, G. and di Serego Alighieri, S. 1991, A. J., **101**, 1647.

Zepf, S.E. *et al.* 2008, Ap. J., **683**, L139.

Infrared

White, D.A. *et al.* 1991, Ap. J., **375**, 35.

NGC 4472/M 49

X-ray

Biller, B.A. *et al.* 2004, Ap. J., **613**, 238.

Forman, W. *et al.* 1993, Ap. J., **418**, L55.

Optical

McNamara, B. *et al.* 1994, A. J., **108**, 844.

Zepf, S.E. *et al.* 2008, Ap. J., **683**, L139.

Radio

Henning, P.A., Sancisi, R. and McNamara, B.R. 1993, Astron. Ap., **268**, 536.

NGC 4676

Multiwavelength

Hibbard, J. and van Gorkom, J. 1996, A. J., **111**, 655.

Optical

Toomre, A. and Toomre, J. 1972, Ap. J., **178**, 623.

Infrared

Rossa, J. *et al.* 2007, A. J., **134**, 2124.

NGC 5194/M 51

Multiwavelength

Rand, R.J., Kulkarni, S.R., and Rice, W. 1992, Ap. J., **390**, 66.

X-ray

Ehle, M., Pietsch, W. and Beck, R. 1995, Astron. Ap., **295**, 289.

Marston, A.P. *et al.* 1995, Ap. J., **438**, 663.

Terashima, Y. and Wilson, A.S. 2001, Ap. J. **560**, 139.

Ultraviolet

Maoz, D. *et al.* 1996, Ap. J. Suppl., **107**, 215.

Infrared

Hippelein, H. *et al.* 1996a, Astron. Ap., **315**, L79.

Hippelein, H. *et al.* 1996b, Astron. Ap., **315**, L82.

Rice, W. 1993, A. J., **105**, 67.

Sauvage, M. *et al.* 1996, Astron. Ap., **315**, L89.

Radio

Condon, J.J. 1987, Ap. J. Suppl., **65**, 485.

Crane, P.C. and van der Hulst, J.M. 1992, A. J., **103**, 1146.

Neininger, N. 1992, Astron. Ap., **263**, 30.

Rand, R.J. and Kulkarni, S.R. 1990, Ap. J., **349**, L43.

Rots, A.H. *et al.* 1990, A. J., **100**, 387.

Vogel, S.N., Kulkarni, S.R. and Scoville, N.Z. 1988, Nature, **334**, 402.

NGC 520

Multiwavelength

Stanford, S.A. and Balcells, M. 1991, Ap. J., **370**, 118.

Hibbard, J. and van Gorkom, J. 1996, A. J., **111**, 655.

X-ray

Read, A.M. 2005, M.N.R.A.S., **359**, 455.

Infrared

Bushouse, H., Stanford, S.A. and Balcells, M. 1992, Ap. J. Suppl., **79**, 213.

Rossa, J. *et al.* 2007, A. J., **134**, 2124.

Stanford, S.A. and Balcells, M. 1990, Ap. J., **355**, 59.

Radio

Hibbard, J.E., Vacca, W.D. and Yun, M.S. 2000, A. J. **119**, 1130.

NGC 1275

Multiwavelength

McNamara, B.R., O'Connell, R.W. and Sarazin, C.L. 1996, A. J., **112**, 91.

Gamma ray

Abdo, A.A. *et al.* 2009, Ap. J., **669**, 31.

X-ray

Böhringer, H. *et al.* 1993, M.N.R.A.S., **264**, L25.

Ultraviolet

Maoz, D. *et al.* 1995, Ap. J., **440**, 91.

Smith, E.P. *et al.* 1992, Ap. J., **395**, L49.

Optical

Cowie, L.L. *et al.* 1983, Ap. J., **272**, 29.

Goudfrooij, P. *et al.* 1994a, Astron. Ap. Suppl., **104**, 179.

Goudfrooij, P. *et al.* 1994b, Astron. Ap. Suppl., **105**, 341.

Holtzman, J.A. *et al.* 1992, A. J., **103**, 691.

Radio

Inoue, M.Y. *et al.* 1996, A.J., **111**, 1852.

Lazareff, B. *et al.* 1989, Ap. J., **336**, L13.

Pedlar, A. *et al.* 1990, M.N.R.A.S., **246**, 477.

NGC 1316

Multiwavelength

Fabbiano, G., Fassnacht, C. and Trinchieri, G. 1994, Ap. J., **434**, 67.

Kim, D.-W., Fabbiano, G. and Mackie, G. 1998, Ap. J., **497**, 699.

Mackie, G. and Fabbiano, G. 1998, A. J., **115**, 514.

X-ray

Feigelson, E.D. *et al.* 1995, Ap. J., **449**, L149.

Kim, D.-W. and Fabbiano, G. 2003, Ap. J., **586**, 826.

Optical

Grillmair, C.J. *et al.* 1999, A. J., **117**, 167.

Schweizer, F. 1980, Ap. J., **237**, 303.

Schweizer, F. 1981, Ap. J., **246**, 722.

Shaya, E.J. *et al.* 1996, A. J., **111**, 2212.

Radio
Birkinshaw, M. and Davies, R.L. 1985, Ap. J., **291**, 32.
Ekers, R.D. *et al.* 1983, Astron. Ap., **127**, 361.
Fomalont, E.B. *et al.* 1989, Ap. J., **346**, L17.
Geldzahler, B.J. and Fomalont, E.B. 1984, A. J., **89**, 1650.
Horellou, C. *et al.* 2001, Astron. Ap., **376**, 837.

NGC 4038/9

X-ray
Fabbiano, G., Schweizer, F. and Mackie, G. 1997, Ap. J., **478**, 542.
Fabbiano, G. *et al.* 2003, Ap. J., **598**, 272.
Read, A.M., Ponman, T.J. and Wolstencroft, R.D. 1995, M.N.R.A.S., **277**, 397.

Ultraviolet
Hibbard, J.E. *et al.* 2005, Ap. J., **619**, L87.

Optical
Amram, P. *et al.* 1992, Astron. Ap., **266**, 106.
Whitmore, B.C. and Schweizer, F. 1995, Ap. J., **109**, 960.

Infrared
Rossa, J. *et al.* 2007, A. J., **134**, 2124.
Vigroux, L. *et al.* 1996, Astron. Ap., **315**, L93.

Radio
Condon, J.J. 1987, Ap. J. Suppl., **65**, 485.
Hibbard, J.E. *et al.* 2001, A. J., **122**, 2969.
Hummel, E. and van der Hulst, J.M. 1986, Astron. Ap., **155**, 151.
Stanford, S.A. *et al.* 1990, Ap. J., **349**, 492.

NGC 7252

Multiwavelength
Hibbard, J.E. *et al.* 1994, A. J., **107**, 67.
Hibbard, J.E. and van Gorkom, J.H. 1996, A. J., **111**, 655.

Optical
Schweizer, F. 1982, Ap. J., **252**, 455.
Whitmore, B.C. *et al.* 1993, A. J., **106**, 1354.

Radio
Dupraz, C. *et al.* 1990, Astron. Ap., **228**, L5.

NGC 253

Gamma ray
Itoh, C. *et al.* 2002, Astron. Ap., **396**, L1.

X-ray
Pietsch, W., 1994, in *Panchromatic View of Galaxies*, eds. G. Hensler, Ch. Theis, J.S. Gallagher (Edition Frontiers), p. 137.
Vogler, A. and Pietsch, W., 1999, Astron. Ap., **342**, 101.
Weaver, K.A. *et al.* 2002, Ap. J., **576**, 19.

Infrared
Rice, W. 1993, A. J., **105**, 67.
Sams, B.J. *et al.* 1994, Ap. J., **430**, L33.

Radio
Canzian, B., Mundy, L.G., and Scoville, N.Z. 1988, Ap. J., **333**, 157.
Condon, J.J. 1987, Ap. J. Suppl., **65**, 485.
Koribalski, B., Whiteoak, J.B. and Houghton, S. 1995, Publ. Astron. Soc. Aust., **12**, 20.
Puche, D., Carignan, C., and van Gorkom, J.H. 1991, A. J., **101**, 456.
Scoville, N.Z. *et al.* 1985, Ap. J., **289**, 129.
Ulvestad, J.S. and Antonucci, R.R.J. 1997, Ap. J., **488**, 621.

NGC 3034/M 82

Multiwavelength
Golla, G., Allen, M.L. and Kronberg, P.P. 1996, Ap. J., **473**, 244.
Yao, L. 2009, Ap. J., **705**, 766.

X-ray
Bregman, J.N., Schulman, E. and Tomisaka, K. 1995, Ap. J., **439**, 155.
Watson, M.G., Stanger, V. and Griffiths, R.E. 1984, Ap. J., **286**, 144.

Infrared
Achtermann, J.M. and Lacy, J.H. 1995, Ap. J., **439**, 163.
McLeod, K.K. *et al.* 1993, Ap.J., **412**, 111.
Rice, W. *et al.* 1988, Ap. J. Suppl., **68**, 91.

Radio
Condon, J.J. 1987, Ap. J. Suppl., **65**, 485.
Matsushita, S. *et al.* 2000, Ap. J., **545**, L107.
Seaquist, E.R. *et al.* 1996, Ap. J., **465**, 691.
Seaquist, E.R., Lee, S.W. and Moriarty-Schieven, G.H. 2006, Ap. J., **638**, 148.
Tilanus, R.P.J. *et al.* 1991, Ap. J., **376**, 500.
Yun, M.S., Ho, P.T.P., and Lo, K.Y. 1994, Nature, **372**, 530.

NGC 5236/M 83

X-ray
Ehle, M. *et al.* 1998, Astron. Ap., **329**, 39.
Soria, R. and Wu, K. 2002, Astron. Ap., **384**, 99.

Infrared
Dong, H. *et al.* 2008, A. J., **136**, 479.
Rice, W. 1993, A. J., **105**, 67.
Rouan, D. *et al.* 1996, Astron. Ap., **315**, L141.

Radio
Allen, R.J., Atherton, P.D., and Tilanus, R.P.J. 1986, Nature, **319**, 296.
Condon, J.J. 1987, Ap. J. Suppl., **65**, 485.
Handa, T. *et al.* 1990, Publ. Astron. Soc. Japan, **42**, 1.
Kenney, J.D.P. and Lord, S.D 1991, Ap. J., **381**, 118.
Lord, S.D. and Kenney, J.D.P. 1991, Ap. J., **381**, 130.
Muraoka, K. *et al.* 2009, Ap. J., **706**, 1213.
Tilanus, R.P.J. and Allen R.J. 1993, Astron. Ap., **274**, 707.

NGC 1068/M 77

X-ray
Ogle, P.M. *et al.* 2003, Astron. Ap., **402**, 849.
Wilson, A.S. *et al.* 1992, Ap. J., **391**, L75.
Young, A.J., Wilson, A.S. and Shopbell, P.L. 2001, Ap. J., **556**, 6.

Optical
Antonucci, R.R.J. and Miller, J.S. 1985, Ap. J., **297**, 621.
Evans, I.N. *et al.* 1991, Ap. J., **369**, L27.

Radio
Condon, J.J. 1987, Ap. J. Suppl., **65**, 485.
Gallimore, J.F., Baum, S.A. and O'Dea, C.P. 2004, Ap. J., **613**, 794.
Planesas, P., Scoville, N. and Myers, S.T. 1991, Ap. J., **369**, 364.
Wilson, A.S. and Ulvestad, J.S. 1983, Ap. J., **275**, 8.

NGC 1365

Multiwavelength
Lindblad, P.O. 1999, Astron. Ap. Review, **9**, 221. Also
http://nedwww.ipac.caltech.edu/level5/Lindblad/
    frames.html

X-ray
Turner, T.J., Urry, C.M. and Mushotzky, R.F. 1993, Ap. J., **418**, 653.
Risaliti, G. *et al.* 2005, Ap. J., **623**, L93.

Optical
Osterbrock, D.E. 1981, Ap. J., **249**, 462.

Radio
Beck, R. *et al.* 2005, Astron. Ap., **444**, 739.
Condon, J.J. 1987, Ap. J. Suppl., **65**, 485.
Jorsater, S. and van Moorsel, G.A. 1995, A. J., **110**, 2037.

NGC 3031/M 81

Multiwavelength
Gordon, K.D. *et al.* 2004, Ap. J. Suppl., **154**, 215.
Ho, L.C., Filippenko, A.V. and Sargent, W.L.W. 1996, Ap. J., **462**, 183.
Kendall, S. *et al.* 2008, M.N.R.A.S., **387**, 1007.

X-ray
Swartz, D.A. *et al.* 2002, Ap. J., **574**, 382.

Ultraviolet
Hill, J.K. *et al.* 1992, Ap. J., **395**, L37.
O'Connell, R.W. *et al.* 1992, Ap. J., **395**, L45.

Optical
Aldering, G.S. and Bothun, G.D. 1991, P.A.S.P., **103**, 1296.

Infrared
Rice, W. 1993, A. J., **105**, 67.

Radio
Adler, D.S. and Westpfahl, D.J. 1996, A. J., **111**, 735.
Condon, J.J. 1987, Ap. J. Suppl., **65**, 485.
Schoofs, S. 1992, Diploma thesis, University of Bonn.
Yun, M.S., Ho, P.T.P., and Lo, K.Y. 1994, Nature, **372**, 530.

NGC 4258/M 106

X-ray
Cecil, G., Wilson, A.S. and De Pree, C. 1995, Ap. J., **440**, 181.
Pietsch, W. *et al.* 1994, Astron. Ap., **284**, 386.
Wilson, A.S., Yang, Y. and Cecil, G. 2001, Ap. J., **560**, 689.

Optical
Ford, H.C. *et al.* 1986, Ap. J., **311**, L7.
Siopis, C. *et al.* 2009, Ap. J., **693**, 946.

Infrared
Rice, W. 1993, A. J., **105**, 67.

Radio
Cox, P. and Downes, D. 1996, Ap. J., **473**, 219.
Greenhill, L.J. *et al.* 1995, Ap. J., **440**, 619.
Hummel, E., Krause, M., and Lesch, H. 1989, Astron. Ap., **211**, 266.
Humphreys, E.M.L. *et al.* 2008, Ap. J., **672**, 800.
van Albada, G.D. 1980, Astron. Ap., **90**, 123.
van Albada, G.D. and van der Hulst, J.M. 1982, Astron. Ap., **115**, 263.

3C 273

Multiwavelength
Courvoisier, T.J.-L. 1998, Astron. and Ap. Review, **9**, 1. Also
http://nedwww.ipac.caltech.edu/level5/Courvoisier/frames.html

X-ray
Espaillat, C. *et al.* 2008, Ap. J., **679**, 182.
Jester, S. *et al.* 2006, Ap. J., **648**, 900.
Marshall, H.L. 2001, Ap. J., **549**, 167.

Optical
Blandford, R.D. and McKee, C.F. 1982, Ap. J., **255**, 419.
Hutchings, J.B. and Neff, S.G. 1991, P.A.S.P., **103**, 26.
Jester, S. *et al.* 2001, Astron. Ap., **373**, 447.
Sparks, W.B., Biretta, J.A. and Macchetto, F. 1993, Ap. J. Suppl., **90**, 909.

Radio
Cohen, M.H. *et al.* 1983, Ap. J., **272**, 383.
Conway, R.G. *et al.* 1993, Astron. Ap., **267**, 347.
Vermeulen, R.C. and Cohen, M.H. 1994, Ap. J., **430**, 467.

NGC 4486/M 87

Multiwavelength
Biretta, J.A. 1993, in *Astrophysical Jets*, eds. D. Burgarella, M. Livio and C.P. O'Dea, (Cambridge University Press, Cambridge). Also
http://nedwww.ipac.caltech.edu/level5/Biretta/frames.html

Gamma ray
Abdo, A.A. *et al.* 2009, Ap. J., **707**, 55.
Aharonian, F. *et al.* 2003, Astron. Ap., **403**, L1.

X-ray
Forman, W. *et al.* 2007, Ap. J., **665**, 1057.
Wilson, A.S. and Yang, Y. 2002, Ap. J., **568**, 133.

Ultraviolet
Boksenberg, A. *et al.* 1992, Astron. Ap., **261**, 393.
Maoz, D. *et al.* 1996, Ap. J. Suppl., **107**, 215.
Madrid, J.P. 2009, A. J., **137**, 3864.

Optical
Davis, L.E. *et al.* 1985, A. J., **90**, 169.
Ford, H.C. *et al.* 1994, Ap. J., **435**, L27.
Goudfrooij, P. *et al.* 1994a, Astron. Ap. Suppl., **104**, 179.
Sargent, W.L.W. *et al.* 1978, Ap. J., **221**, 731.
Sparks, W.B., Ford, H.C. and Kinney, A.L. 1993, Ap. J., **413**, 531.
Sparks, W.B., Biretta, J.A. and Macchetto, F. 1993, Ap. J. Suppl., **90**, 909.

Radio
Biretta, J.A., Zhou, F. and Owen, F.N. 1995, Ap. J., **447**, 582.
Birkinshaw, M. and Davies, R.L. 1985, Ap. J., **291**, 32.
Junor, W., Biretta, J.A. and Livio, M. 1999, Nature, **401**, 891.

NGC 4594/M 104

X-ray
Fabbiano, G. and Juda, J.Z. 1997, Ap. J., **476**, 666.
Pellegrini, S. *et al.* 2002, Astron. Ap., **383**, 1.

Radio
Condon, J.J. 1987, Ap. J. Suppl., **65**, 485.
Krause, M., Wielebinski, R. and Dumke, M. 2006, Astron. Ap., **448**, 133.

NGC 5128

Multiwavelength
Israel, F.P. 1998, Astron. Ap. Review, **8**, 237. Also http://nedwww.ipac.caltech.edu/level5/March01/Israel/frames.html

Gamma ray
Aharonian, F. *et al.* 2009, Ap. J., **695**, L40.

X-ray
Döbereiner, S., *et al.* 1996, Ap. J., **470**, L15.
Karovska, M. *et al.* 2002, Ap. J., **577**, 114.
Kraft, R.P. *et al.* 2000, Ap. J., **531**, L9.

Optical
Lauberts, A. and Valentijn, E.A. 1989, *The Surface Photometry Catalogue of the ESO-UPPSALA Galaxies* (ESO, Garching Bei Munchen).
Marconi, A. *et al.* 2001, Ap. J., **549**, 915.
Morganti, R. *et al.* 1991, M.N.R.A.S., **249**, 91.
Rejkuba, M. *et al.* 2005, Ap. J., **631**, 262.
Schreier, E.J. *et al.* 1996, Ap. J., **459**, 535.

Radio
Condon, J.J. *et al.* 1996, Ap. J. Suppl., **103**, 81.
Burns, J.O., Feigelson, E.D. and Schreier, E.J. 1983, Ap. J., **273**, 128.
Jauncey, D. *et al.* 1995, Proc. Nat. Acad. Sci. USA, **92**, 11368.
Jones, D.L. *et al.* 1996, Ap. J., **466**, L63.

Junkes, N. *et al.* 1993, Astron. Ap., **269**, 29.
Meier, D.L. *et al.* 1989, A. J., **98**, 27.
Meier, D.L. *et al.* 1993, in *Sub-Arcsecond Radio Astronomy*, ed. Davis & Booth, (Cambridge University Press, Cambridge), p. 201.
Quillen, A.C. *et al.* 1992, Ap. J., **391**, 121.
Rydbeck, G. *et al.* 1993, Astron. Ap., **270**, L13.
Tingay, S.J. *et al.* 1994, Aust. J. Physics, 47, 619.
Tingay, S.J. *et al.* 1996, in *Energy Transport in Radio Galaxies and Quasars*, ASP conference series vol. 100, Eds. P.E. Hardee, A.H. Bridle, and J.A. Zensus (ASP, San Francisco), p. 215.
Tingay, S.J. *et al.* 1998, A. J., **122**, 1697.
van Gorkom, J.H. *et al.* 1990, A. J., **99**, 1781.

A1795 #1

Multiwavelength
Ge, J.P. and Owen, F.N. 1993, A. J., **105**, 778.

X-ray
Briel, U.G. and Henry, J.P. 1996, Ap. J., **472**, 131.
Fabian, A.C. *et al.* 2001, M.N.R.A.S., **321**, L33.

Ultraviolet
Smith, E.P. *et al.* 1997, Ap. J., **478**, 516.

Optical
Cowie, L.L. *et al.* 1983, Ap. J., **272**, 29.
McNamara, B.R. *et al.* 1996a, Ap. J., **466**, L9.
McNamara, B.R. *et al.* 1996b, Ap. J., **469**, 69.
Pinkney, J. *et al.* 1996, Ap. J., **468**, L13.

Radio
Liuzzo, E. *et al.* 2009, Astron. Ap., **501**, 933.

ARP 220

Optical
Sanders, D.B. *et al.* 1988, Ap. J., **325**, 74.
Shaya, E.J. *et al.* 1994, A. J., **107**, 1675.

Infrared
Norris, R.P. 1985, M.N.R.A.S., **216**, 701.
Smith, C.H., Aitken, D.K., and Roche, P.F. 1989, M.N.R.A.S., **241**, 425.

Radio
Lonsdale, C.J. *et al.* 2006, Ap. J., **647**, 185.
Norris, R.P. 1988, M.N.R.A.S., **230**, 345.
Scoville, N.Z. *et al.* 1991, Ap. J., **366**, L5.

CYGNUS A

Multiwavelength
Carilli, C.L. and Harris, D.E. (eds) 1996, *Cygnus A – Study of a Radio Galaxy*, (Cambridge University Press, Cambridge).
Carilli, C.L. and Barthel, P.D. 1996, Astron. Ap. Review, 7, 1. Also http://nedwww.ipac.caltech.edu/level5/Carilli/frames.html

X-ray
Carilli, C.L., Perley, R.A. and Harris, D.E. 1994, M.N.R.A.S., **270**, 173.

Harris, D.E., Carilli, C.L. and Perley, R.A. 1994, Nature, **367**, 713.

Steenbrugge, K.C., Blundell, K.M. and Duffy, P. 2008, M.N.R.A.S., **388**, 1465.

Wilson, A.S., Young, A.J. and Shopbell, P.L. 2000, Ap. J. **544**, L27.

Young, A.J. *et al.* 2002, Ap. J., **564**, 176.

Radio

Carilli, C.L. 1989, Ph.D. thesis (MIT).

Carilli, C.L., Bartel, N. and Linfield, R.P. 1991, A. J., **102**, 1691.

Perley, R.A., Dreher, J.W. and Cowan, J.J. 1984, Ap. J., **285**, L35.

Acknowledgements and permissions

Many thanks to the authors and publishers who have granted permission to reproduce their data.

All-sky Galaxy images are courtesy of the Astrophysics Data Facility at NASA's Goddard Space Flight Center; http://mwmw.gsfc.nasa.gov/

This work has made use of the SIMBAD database (http://simbad.u-strasbg.fr), operated at CDS, Strasbourg, France.

This atlas also benefited from the NASA/IPAC Extragalactic Database (NED http://nedwww.ipac.caltech.edu) which is operated by the Jet Propulsion Laboratory, California Institute of Technology, under contract with the National Aeronautics and Space Administration.

Every effort has been made to secure necessary permissions to reproduce copyright material in this work, though in some cases it has proved impossible to trace copyright holders. If any omissions are brought to notice, appropriate acknowledgements on reprinting or in any subsequent edition will be given.

QUOTATIONS USED IN THE TEXT

Chapter 1

W. Saslaw, from Saslaw, W.C. 1985, *Gravitational Physics of Stellar and Galactic Systems*, Chapter 21 (Cambridge University Press, Cambridge), p. 157.

J.F.W. Herschel, from Herschel, J.F.W. 1847, "Results of astronomical observations made during the years 1834, 5, 6, 7, 8, at the Cape of Good Hope; being the completion of a telescopic survey of the whole surface of the visible heavens, commenced in 1825" (Smith, Elder and Co., London).

D. Silvan Evans, from Evans, D. Silvan. 1870, Nature, *The Milky Way*, vol. 3, p. 167. Reprinted by permission from Macmillan Publishers Ltd: *Nature*, vol. 3, p. 167, copyright (1870).

Henrietta Leavitt, from Leavitt, H.S. 1908, Annals of Harvard College Observatory, **60**, 87.

Allan Sandage, from Sandage, A. 1961, *The Hubble Atlas of Galaxies* (Carnegie Institution, Washington D.C.). Reprinted with permission.

James B. Kaler, from Kaler, James, B. 2000, Astronomy, Sept. Reproduced by permission. ©2000 *Astronomy* magazine, Kalmbach Publishing Co.

Bernard Pagel, from Pagel, B.F. 1985 in *Active Galactic Nuclei*, ed. J.E. Dyson, (Manchester University Press, Manchester), p. 373.

Allan Sandage, from Sandage, A. 2005, Ann. Rev. Astron. Astrophys., **43**, 581. With permission, from the *Annual Review of Astronomy and Astrophysics* by Allan Sandage. ©2005 by Annual Reviews Inc.

Sidney van den Bergh, from van den Bergh, S. 1998, *Galaxy Morphology and Classification* (Cambridge University Press, Cambridge), p. 1.

Harlow Shapley, from Shapley, H. *Galaxies*, 1973 (3rd edition, Harvard University Press, Cambridge), p. 32. Reprinted by permission of the publisher from GALAXIES by Harlow Shapley, revised by Paul Hodge, p. 32, Cambridge, Mass.: Harvard University Press, Copyright ©1943, 1961, 1972 by the President and Fellows of Harvard College.

Chapter 2

Walter Baade, from Baade, W. 1963, *Evolution of Stars and Galaxies* (Harvard University Press, Cambridge), p. 54. Reprinted by permission of the publisher from *Evolution of Stars and Galaxies*, by Walter Baade, p. 54, Cambridge, Mass.: Harvard University Press, Copyright ©1963 by the President and Fellows of Harvard College.

Vera Rubin, from Lightman, A. and Brawer, R. 1992, *Origins: The Lives and Worlds of Modern Cosmologists* (Harvard University Press, Cambridge) p. 297. Reprinted by permission of the publisher from Origins: The Lives and Worlds of Modern Cosmologists, by Alan Lightman and Roberta Brawer, p. 297, Cambridge, Mass.: Harvard University Press, Copyright ©1990 by Alan Lightman and Roberta Brawer.

Ken Freeman from Freeman, K.C. 1970, Ap. J., **160**, 811. Reproduced by permission of the AAS.

IMAGE, TABLE AND FIGURE PERMISSIONS

Figure 1.2: F. Winkler/Middlebury College, the MCELS Team and NOAO/AURA/NSF. The MCELS project has been supported in part by NSF grants AST-9540747 and AST-0307613, and through the generous support of the Dean B. McLaughlin Fund at the University of Michigan. The National Optical Astronomy Observatory (NOAO) is operated by the Association of Universities for Research in Astronomy Inc. (AURA), under a cooperative agreement with the National Science Foundation (NSF).

Figure 1.9: NGC 598/M 33. The galaxy was observed with the KPNO 4 m and MosaicI camera by Phil Massey (Lowell Obs.) and Shadrian Holmes (Univ. Texas). Credit: P. Massey and the Local Group Survey Team.

Figure 1.19: The Magellanic Cloud-type ImIV-V galaxy NGC 6822. The image was taken by the CTIO 4 m with the MosaicII camera by Knut Olsen and Chris Smith (CTIO). Credit: P. Massey and the Local Group Survey Team.

Figure 1.24: The Large Magellanic Cloud. C. Smith, S. Points, the MCELS Team and NOAO/AURA/NSF. The MCELS project has been supported in part by NSF grants AST-9540747 and AST-0307613, and through the generous support of the Dean B. McLaughlin Fund at the University of Michigan. The National Optical Astronomy Observatory (NOAO) is operated by the Association of Universities for Research in Astronomy

Inc. (AURA), under a cooperative agreement with the National Science Foundation (NSF).

Table 3: Original table from Gallagher, J. and Fabbiano, G. 1990, in *Windows on Galaxies*, eds. Fabbiano, G., Gallagher, J.S., and Renzini, A. (Kluwer, Dordrecht), p. 1. With kind permission of Springer Science and Business Media.

Figure 2.8: Optical spectra of dwarf stars. Credit: Supplied by R. Pogge. Figure courtesy of G. Jacoby/NOAO/AURA/NSF. Reproduced by permission of the AAS.

Figure 2.9: The Large Magellanic Cloud in Hα. C. Smith, S. Points, the MCELS Team and NOAO/AURA/NSF. The MCELS project has been supported in part by NSF grants AST-9540747 and AST-0307613, and through the generous support of the Dean B. McLaughlin Fund at the University of Michigan. The National Optical Astronomy Observatory (NOAO) is operated by the Association of Universities for Research in Astronomy Inc. (AURA), under a cooperative agreement with the National Science Foundation (NSF).

Figure 2.13: Radio/sub-mm/far-IR spectrum of NGC 3034/M 82. Credit: With permission, from the *Annual Review of Astronomy and Astrophysics*, Volume 30 ©1992 by Annual Reviews www.annualreviews.org.

Figure 2.16: Stellar density map of the Andromeda–Triangulum region. Credit: Alan McConnachie. The images of M 31 and M 33 courtesy T. A. Rector and M. Hanna. Reprinted by permission from Macmillan Publishers Ltd: *Nature*, vol. 461, p. 66, copyright (2009).

Figure 2.17: JCMT SCUBA 850 μm image of the Hubble Deep Field – North. Data from Hughes *et al.* (1998). Reprinted by permission from Macmillan Publishers Ltd: *Nature*, vol. 394, p. 241, copyright (1998).

Figure 2.21: 1.49 GHz radio power versus IR luminosity for IRAS galaxies. Credit: Figure from Eli Dwek – original form in Dwek and Barker (2002). Reproduced by permission of the AAS.

Figure 3.8: The Galaxy: Optical – this image is from Lund Observatory. Kindly provided by Prof. L. Lindegren, Director, Lund Observatory.

Figure 3.9: The Galaxy: Optical – Hα at 6563 Å. Credit: Permission granted by D. Finkbeiner. Original data from WHAM (funded primarily through the National Science Foundation, NSF), VTSS and SHASSA (both supported by the NSF). The SHASSA data is ©Las Cumbres Observatory, Inc.

Figure 4.10: NGC 224/M 31: Radio CO 2.6 mm. Copyright: ©MPIfR Bonn/IRAM (Ch. Nieten, N. Neininger, M. Guelin *et al.*).

Figure 4.11: NGC 224/M 31: Radio 6 cm continuum and magnetic field. Copyright: ©MPIfR Bonn (R. Beck, E.M. Berkhuijsen, P. Hoernes).

Figure 4.35: NGC 891: Optical V. Copyright WIYN Consortium Inc., all rights reserved.

Figure 4.99: NGC 1316: Optical Hα + [N II]. Reproduced by permission of the AAS.

Figure 4.123: NGC 3034/M 82: Optical B, V, Hα. ©Subaru Telescope, National Astronomical Observatory of Japan (NAOJ) and is reproduced with permission.

Figure 4.135: NGC 5236/M 83: Radio HI. The image is ©CSIRO Australia 2004.

Figure 4.195: Cygnus A: Near-IR K', Reproduced with permission of Lawrence Livermore National Laboratory/ W.M. Keck Observatory.

Figure A.5: Parkes 64 m radio telescope. © Seth Shostak.

Galaxy index

Index